DESIGN FOR EMERGENCE

Frontiers in Artificial Intelligence and Applications

Volume 153

Published in the subseries

Dissertations in Artificial Intelligence
Under the Editorship of the ECCAI Dissertation Board

Recently published in this series

Vol. 152. T.M. van Engers (Ed.), Legal Knowledge and Information Systems – JURIX 2006: The Nineteenth Annual Conference
Vol. 151. R. Mizoguchi et al. (Eds.), Learning by Effective Utilization of Technologies: Facilitating Intercultural Understanding
Vol. 150. B. Bennett and C. Fellbaum (Eds.), Formal Ontology in Information Systems – Proceedings of the Fourth International Conference (FOIS 2006)
Vol. 149. X.F. Zha and R.J. Howlett (Eds.), Integrated Intelligent Systems for Engineering Design
Vol. 148. K. Kersting, An Inductive Logic Programming Approach to Statistical Relational Learning
Vol. 147. H. Fujita and M. Mejri (Eds.), New Trends in Software Methodologies, Tools and Techniques – Proceedings of the fifth SoMeT_06
Vol. 146. M. Polit et al. (Eds.), Artificial Intelligence Research and Development
Vol. 145. A.J. Knobbe, Multi-Relational Data Mining
Vol. 144. P.E. Dunne and T.J.M. Bench-Capon (Eds.), Computational Models of Argument – Proceedings of COMMA 2006
Vol. 143. P. Ghodous et al. (Eds.), Leading the Web in Concurrent Engineering – Next Generation Concurrent Engineering
Vol. 142. L. Penserini et al. (Eds.), STAIRS 2006 – Proceedings of the Third Starting AI Researchers' Symposium
Vol. 141. G. Brewka et al. (Eds.), ECAI 2006 – 17th European Conference on Artificial Intelligence
Vol. 140. E. Tyugu and T. Yamaguchi (Eds.), Knowledge-Based Software Engineering – Proceedings of the Seventh Joint Conference on Knowledge-Based Software Engineering

ISSN 0922-6389

Design for Emergence
Collaborative Social Play with Online and Location-Based Media

Yanna Vogiazou
*Lecturer in Interaction Design, Goldsmiths College, University of London
Knowledge Media Institute, The Open University*

Press

Amsterdam • Berlin • Oxford • Tokyo • Washington, DC

© 2007 The author and IOS Press.

All rights reserved. No part of this book may be reproduced, stored in a retrieval system, or transmitted, in any form or by any means, without prior written permission from the publisher.

ISBN 978-1-58603-701-7
Library of Congress Control Number: 2006938567

Publisher
IOS Press
Nieuwe Hemweg 6B
1013 BG Amsterdam
Netherlands
fax: +31 20 687 0019
e-mail: order@iospress.nl

Distributor in the UK and Ireland
Gazelle Books Services Ltd.
White Cross Mills
Hightown
Lancaster LA1 4XS
United Kingdom
fax: +44 1524 63232
e-mail: sales@gazellebooks.co.uk

Distributor in the USA and Canada
IOS Press, Inc.
4502 Rachael Manor Drive
Fairfax, VA 22032
USA
fax: +1 703 323 3668
e-mail: iosbooks@iospress.com

LEGAL NOTICE

The publisher is not responsible for the use which might be made of the following information.

PRINTED IN THE NETHERLANDS

To my mother, Svetlana

Foreword

Design for emergence investigates spontaneous, unpredictable uses of technology that are driven by social contexts and collaborative processes, based on our ability to communicate our presence, both virtual and physical, in symbolic ways. In light of the fact that social dynamics and unexpected uses of technology can inspire innovation, this book proposes a research model of *design for emergence*, focusing on emergent phenomena as part of an iterative design process. By providing playful, technology-mediated experiences with minimal structure, unpredictable user behaviours can emerge through exploration, resulting in a richer and more complex, social experience. The research methodology is practice-based; two interactive prototypes were designed, implemented and evaluated in different contexts: an online multiplayer BumperCar game and a wireless, location-based urban game of 'tag', called CitiTag. User studies showed that collaborative, spontaneous play can enhance the sense of social participation in a group activity. Collective and individual behaviours and creative uses of technology emerged from a simply designed application based on symbolic presence, both in the virtual and the physical world.

CitiTag experiments showed that virtual elements in a mixed reality game can instigate novel experiences in the context of our everyday physical and social environment, with often unexpected results. The observed emergent behaviours are personal and collective extensions of the virtual experience in the real world. The book concludes with a positive view of ubiquitous and social computing, in which the virtual world becomes a 'first class citizen' rather than a substitute for the real world, creating new situations and engaging experiences in the setting of our daily life that were not possible before.

Keywords: spontaneous play, collaboration, emergence, presence, ubiquitous, mixed reality, interaction design.

Contents

Guide:

▥ = Storyboard, idea generation

📖 = Background, literature review

◎ = Conclusions, reflection

🐚 = Design, prototype description

⊥ = Theoretical foundation, principles

➡ = Findings

𝐴 = Methods, processes

👁 = Observational studies

****

FOREWORD	vii
1. INTRODUCTION	1
1.1 Researching Spontaneous Collaborative Play	1
1.2 Thesis Structure	2

PART I. FOUNDATIONS: SOCIAL PLAY AS A DESIGN FRAMEWORK FOR EMERGENCE

2. EXPLORING PRESENCE AND SOCIAL PLAY 📖	7
2.1 Defining Presence Based Group Play	7
2.1.1 What Is Presence?	8
2.1.2 Why Play?	14
2.1.3 Crowd Behaviour and Group Dynamics	17
2.1.4 Emergent Self-Organisation	21
2.2 The Challenge of Mixed Reality Collective Experiences	23
2.2.1 Participant Observation in Flash Mobs	24
2.2.2 Wireless Location-Based Multiplayer Games	27
2.2.3 Location-Based Social Software	30
2.2.4 A Categorization of Ubiquitous Social Experiences	33
2.2.5 Research Framework	38

3. DESIGN FOR EMERGENCE 41
   3.1 A Model for Design for Emergence 41
   3.2 Design Principles 45
       3.2.1 Presence Is Symbolic 45
       3.2.2 Large Scale Is Important for Emergent Interaction 47
       3.2.3 Keep the Design Lightweight 48
       3.2.4 By Employing Affordances Users Understand and Can Extend the Design 49

4. EARLY DESIGN SKETCHES FOR DESIGN FOR EMERGENCE 51
   4.1 Online Games 52
   4.2 Pixeltag: A Mobile Game 55
   4.3 Mixed Reality Games 59
   4.4 Final Thoughts Towards a Playground Social Game 65

PART II. ONLINE CASE STUDY: EXPERIMENTS WITH A MULTIPLAYER BUMPER CAR GAME

5. PLAYGROUND INTERACTION ONLINE: THE BUMPER CAR GAME 69
   5.1 The Idea 69
   5.2 Storyboards and Variations 70
   5.3 Technical Limitations and Design Considerations 72
   5.4 BumperCar Design and Experiments 73
       5.4.1 The Game Design 73
       5.4.2 Experimental Design 76
       5.4.3 Analysis Method 80
   5.5 Findings 82
       5.5.1 Emergence 82
       5.5.2 Spontaneous Collaboration and Group Behaviours 85
       5.5.3 Game Experience 88
       5.5.4 Visual Communication and Design 92
   5.6 Conclusions 97

PART III. DESIGN FOR EMERGENCE IN THE REAL WORLD: EXPERIMENTING WITH A MIXED REALITY URBAN PLAYGROUND

6. CITITAG: URBAN SPACE AS A LARGE GROUP PLAYGROUND   103

    6.1   The Idea   103

    6.2   Storyboards and Scenarios   105
         6.2.1   The Action of Tagging   105
         6.2.2   Group Formations and Swarming   107
         6.2.3   Views   108
         6.2.4   Metaphor   110

    6.3   The CitiTag Game   112
         6.3.1   Limitations and Design Considerations   112
         6.3.2   CitiTag Design   112
         6.3.3   CitiTag System Architecture   115

    6.4   Method: User Studies   116

    6.5   Findings   117
         6.5.1   Game Experience   117
         6.5.2   Emergence   123
         6.5.3   Awareness, Group Belongingness and Collaboration   133
         6.5.4   Usability and Design   136

    6.6   Conclusions and Future Work   138

PART IV. REFLECTIONS ON HOW TO DESIGN FOR EMERGENCE

7. DESIGNING FOR SPONTANEOUS COLLABORATIVE PLAY BASED ON PRESENCE   143

    7.1   Revisiting the Research Framework   143
         7.1.1   Emergence in BumperCars and CitiTag   144
         7.1.2   Emergence as *Experiential Variability*   147
         7.1.3   Emergence Feeding Back into the Design Process   150

    7.2   Guidelines for Design for Emergence for Online and Ubiquitous Multi-User Applications   152

7.3 Future Work with Spontaneous Presence-Based Play   156
    7.3.1 Key Research Questions and Opportunities   156
    7.3.2 UrbanSwarm: A Proposed Example for Further Research   158

7.4 Conclusion: So What?   164

REFERENCES   165

ACKNOWLEDGEMENTS   173

1. Introduction

1.1 Researching spontaneous collaborative play

This thesis is an attempt to address the spontaneous, often unpredictable uses of technology that can foster feelings of social participation. A key idea is that social processes and dynamics advance technological design. Consider the most recent applications that are primarily focused on social interaction: social software (Allen, 2004), match-making and friend-finder websites (e.g. Orkut, Friendster), blogs and messaging systems (e.g. Instant Messaging, SMS, MMS etc). At the same time ubiquitous computing (Weiser, 1991) and mobile technologies have created new opportunities for 'mixed reality' social experiences, where information and virtual narratives can be superimposed on the real world and real people, blurring the boundaries between the physical and virtual world as never before possible. Being in a public space and communicating with a stranger nearby via one's mobile phone is a new concept, associated with a lot of excitement but also with attendant fears.

The thesis investigates an opportunity space created by the ability to manifest our presence, both virtual and physical, in symbolic ways through the use of emerging technologies. One of the first problems the thesis tries to address is whether it is possible to harness the feel-good factor of group presence. In other words, how can the presence of large numbers of people be communicated and what kind of design can enhance the sense of participation and being part of a group. In this context, what is most inspiring is the possibility of emergent, spontaneous interaction that can spring from participating in some kind of large-scale, technology-mediated, social experience. Such a project has many unknowns and risks: there is no way to know in advance what these emergent behaviours and interactions will be like; neither can we ensure that these will occur in the first place. We capitalize on this uncertainty to bring the notion of emergence to the forefront and study it as part of an iterative design process. Creative and social uses of technologies have, to a large extent, formed our experience of the internet as a social medium. Many times these emergent uses, unintended by the designers, computer scientist and engineers have inspired and advanced technological innovation.

This thesis is indicative of a recent trend in Human Computer Interaction (HCI) research in the last few years (Preece et al, 2002), to include work from the humanities disciplines, such as sociology, anthropology, social psychology, arts as well as industrial design. HCI researchers have realised that including a broad spectrum of areas can benefit the field as a whole, addressing the limitations of traditional, task focused HCI problem solving and introducing a variety of research methods and creative design processes for emerging technologies. Already the term 'interaction design' is often used for HCI research to suggest that what is at stake is not only a user's interaction with a computer, but a broader contextual framework, taking into account the environments in which people work, collaborate, learn and play, patterns of interaction and social processes that influence the way technology is being used. Understanding this contextual framework can support the design of successful interactive systems.

The research supporting the thesis has been carried out through practice. It is not possible to research spontaneous participatory play or unintended uses of technology and emergent behaviours without actually implementing a prototype and giving it to real people to experiment with. In contrast to other research in the field though, the prototype is not the final outcome of the study, it is not a tool for a particular task nor does it aim to address a specific design problem. The designs that have been implemented as part of this study are only a means to an end; the aim is not to produce an innovative design as such, but rather to explore user interactions, acquire knowledge about what constitutes an engaging participatory, technology-mediated experience and how emergent and unpredictable behaviours occur. For this reason, there is minimal intervention; the designs attempt to provide just enough context and structure, to encourage experimentation and play. The focus on play is strong throughout the thesis, based on the belief that collaborative, spontaneous play can be a good way to enhance the sense of being part of a group and participation in recreational activities. The thesis is only a starting point, illustrating how emerging technologies, presence, social dynamics and play can combine to create an engaging experience and identifying design guidelines for emergence. Hopefully, it will inspire more, interdisciplinary research in the area. Ubiquitous computing technologies challenge the whole notion of computers as tools or as objects bound to a particular context, but at the same time there are many unknowns concerning how these technologies will blend with the fabric of our everyday life and how we will use them in the future. In this context, this thesis takes a step towards suggesting methods for experimentation and explores the niche of group interaction and social play as valuable inputs for design research.

1.2   Thesis structure

In Part I, Chapter 2 introduces the concept of spontaneous collaborative play by looking at definitions of presence and play. Presence is defined as the feeling of being in touch or 'connected' to other people without necessarily interacting with them. As such, presence can be communicated in rather abstract and symbolic ways. Play is viewed as a fundamental aspect of human nature that can strengthen social bonds among individuals and enhance collaborative practices. We differentiate between play and games and identify our interest in spontaneous, free and collaborative play, which is based on very simple rules and encourages improvisation, much like many children's playground games. In this sense, play is different from the traditional notion of games because it is unrestricted: it does not bind the player to follow a particular route imposed by the design, but instead creates more opportunities for unpredictable and emergent behaviours through play. This is important, because the concept of emergence, as the unexpected, spontaneous individual or collective use of a novel technology is crucial to this thesis, as we will see in the suggested research model for design for emergence in chapter 3. Chapter 2 includes a literature review, drawing from research in different areas: social computing, social psychology, games design and ubiquitous, multi-user applications. The social psychology theories suggest that being part of a crowd or participating in a group anonymously does not necessarily result in negative behaviour (e.g. antisocial crowd behaviours, riots etc) but that the participating individuals are less self-aware and associate more with the group's identity, encouraging behaviour that is normative for the group. These behaviours can be positive and pro-social, engaging participating individuals in playful, social activities (e.g. dancing, singing, playing etc). Next we consider some concepts and examples of emergent, self-organising behaviour and collaborative play that we have found interesting. Chapter 2 provides an overview of the kind of location-based multiuser applications that have been emerging in the last few years, in the realm of both games and social software, with a final categorisation of what we broadly identify as *ubiquitous social experiences*. This is the domain where we aim to place our work, the ubiquitous space of our everyday life, in which people can communicate and collaborate in novel ways.

In chapter 3 we propose our research model to investigate spontaneous, often unintended, emergent uses of these technologies. The chapter introduces the concept of *design for emergence*, as a means of encouraging unpredictable, social uses of technology. We explain why it is important to study emergence and the challenges involved in designing not only for an unknown outcome, but for something that we have no way of ensuring will happen at all. The suggested model for design for emergence captures some aspects of emergent phenomena from unintended uses of technology and brings emergence to the forefront. In this model, 'emergence', as a set of instances of spontaneous individual and group behaviours emerging through the use of a novel technology, is the result of a combination of design and external factors. As such, it can only be 'tested' through the deployment or experiment with an interactive product. The model serves as a guide for the research presented in this thesis and expresses our long-term goal, which is to try and integrate lessons learnt from emergent user behaviours in the design process. After introducing the model for design for emergence, chapter 3 presents four design principles, as key building blocks on which this work is founded: 1) the use of symbolic presence 2) the importance of designing for large scale 3) the principle of lightweight design and 4) the provision of just enough affordances for users to understand and extend the context of the activity they are participating in.

Following from the theoretical framework of the design for emergence model and the design principles of chapter 3, chapter 4 presents a series of design concepts with graphic illustrations for a wide range of applications, which were developed during the conceptual, brainstorming and storyboarding phases of the two research projects described in chapters 5 and 6 respectively. Although none of these concepts was selected for development, they form an important part of the work, highlighting different possibilities and the pros and cons of each design. Several of these concepts are also useful for future work in the area, as the reasons we decided not to develop them were mainly limitations in resources and technologies available at that particular time. For each of these concepts we suggest a speculative model of design for emergence to explain in general terms how we expect the games to facilitate collaborative social play and unintended uses of technologies. First, the model is explored in the light of online game development, so the game concept presented attempts to address the following challenges: how can we design engaging social experiences for large numbers of people online based on the communication of symbolic, abstract presence? What kind of design can encourage spontaneous group behaviours to emerge online through play? We present multiplayer game ideas that blur the boundaries between the virtual and physical worlds, starting with a simple mobile game which is still purely virtual in the sense that it does not create direct links with objects and people in the surrounding environment and then proceeding with more ubiquitous scenarios. These are mixed reality concepts aiming to encourage emergence in the real world, where people's virtual presence can penetrate real, physical presence and unexpected events can happen. Chapter 4 concludes with the rationale of our decision to design and develop the online BumperCar game and the urban playground 'tag' game, called CitiTag, as testbeds for our research.

These two case studies are divided into Part II and Part III respectively: the former focuses purely on online interaction and the latter explores social play in an urban environment using wireless and location-based technology. In Part II, Chapter 5 presents the design, implementation and experimentation with an online 'playground space', a customizable online multiplayer BumperCar game that can be used for different playful activities. We first outline the original scenarios and variations, discussing the underlying design decisions and then present the final prototype and the experimental design. Findings from six online game sessions illustrate that collective behaviours can emerge online, either completely spontaneously with surprising results or within the context of a collaborative group activity. Most importantly, visual communication and awareness of other participants' activity are sufficient for these behaviours to emerge, even without verbal or textual communication. The design for emergence model is revisited in light of these observations. We also identify factors that influenced the participants' experience when playing these games. Other results inform the visual communication and design of these activities, showing how a minimal two-dimensional design with bumper cars can provide enough context and even encourage people to assume personality elements for other players.

Part III comprises the most mature phase of this thesis. Chapter 6 describes the concept of 'playground tag', underlying the design and implementation of the CitiTag game, a wireless location-based multiplayer game for spontaneous play in urban environments. In order to further investigate how complexity can grow out of a simply designed application, this work with the CitiTag game attempts to explore spontaneous emergent individual and group behaviours in the real world, motivated by people's participation in a mixed reality game experience. Again, some of the original storyboards are included to illustrate how the design ideas evolved. Chapter 6 presents findings from two user trials, one with nine people at the Open University campus and one with sixteen at the city centre of Bristol. The studies show that a) the location b) the group dynamics and the social aspect c) using the real world as a game board and d) the match between the virtual, fictional events and the physical, real world interface are all important factors that affect participant experience. The studies also demonstrate how, when people push boundaries on various fronts, collaborative or unexpected behaviours can emerge in the real world, mediated by superimposed, symbolic virtual presence.

Part IV, chapter 7 of this thesis brings our two studies in Parts II and III together, reflecting on the similarities and differences of emergence in our online and mixed reality game studies. We explore and propose a set of guidelines for design for emergence in future social multiuser applications, both online and mediated through ubiquitous and mobile technologies. Then we suggest future research work with a design proposal for UrbanSwarm, a multiplayer game promoting large-scale spontaneous collaborative play that can blend with the fabric of our everyday life. Chapter 7 summarizes key points from the thesis and concludes with thoughts about a research approach in which the virtual world is elevated to 'first class citizenship' by becoming part of our daily reality and affecting our surrounding physical and social environment, rather than being a limited virtual alternative to a real life experience.

Part I

Foundations: Social Play as a Design Framework for Emergence

2. Exploring presence and social play

2.1 Defining presence based group play

Current advances and convergence trends in communication technologies have been changing the ways we communicate with other people. A sense of 'being connected' or 'always in touch' is achieved even when not directly interacting with information devices themselves. In the networked world the 'presence' awareness of other people can be achieved with a range of communication tools, Instant Messaging being a typical example. People manifest their presence to others in various ways and situations when they are not physically co-present, like sending playful text messages while being on the move.

This chapter illustrates that the notions of 'presence' and social play are in fact, intertwined; and interactive media that communicate both are emerging and developing fast. The examples of location-based games and social software in the literature review that follows highlight the exciting opportunities for engaging, social, playful user experiences arising from blending the boundaries between virtual and physical presence.

Of particular interest to this research are group presence and the following question: How can we harness the sense of the simultaneous presence of many people? As Donath (1996) asks: Is there a design that would make palpable the sensation that one was indeed on-line in the company of millions of other people? There is definitely a particular feeling when being part of a crowd, but the ways in which one can get this sense on a large scale in the online and wireless world have not been fully identified.

In addition to this difficult question, we have been investigating the potential of emergent collaborative social behaviours based on group presence. One of the most intriguing contemporary ideas following the publication of Rheingold's 'Smart Mobs' at the end of 2002, is the empowerment of self-organization and large group coordination with mobile technologies. This is exemplified by the spontaneous assemblies of protesters against the WTO in Seattle and protesters in Philippines (Rheingold 2002), which were coordinated via SMS. The thesis does not focus on emergent behaviours in a political context, but rather in the context of collaborative play. We are motivated by the belief that collaborative recreational activities can foster the sense of being part of a group, what we refer to as 'group belongingness'. It is precisely this feeling we aim to harness through the design of engaging technology mediated social experiences.

The next sections illuminate the basic 'ingredients' of emergent, presence-based play: presence, play, group and crowd dynamics. We have considered examples and looked into background research literature, wherever available, for each of these aspects. First, the concept of presence is defined in the next section 2.1.1, as a feeling that can be communicated in symbolic and quite abstract ways in the context of what has been described as *social presence*: the sense of being together or in touch with other people. We then investigate and define 'play' and identify the values of play that are important for this research, discussing the difference between spontaneous, free play and games. The prototypes designed and implemented for this thesis belong to the sphere of spontaneous play, even though they have game-like elements in them. The focus on free play has enabled us to design applications that encourage exploration and experimentation, with often unexpected user behaviours. It is precisely these unpredictable behaviours, particularly the collective ones, that we aim for. We then outline some interesting points from several social psychology theories in section 2.1.3, which help us understand some possible effects of being part of a group or a crowd. Following from the paradigm of the Mexican Wave, a large-scale collective crowd behaviour, section 2.1.4 deals with self-organisation and explains how certain macrobehaviours emerge in different examples.

2.1.1 What is presence?

Presence, both physical and virtual, is an essential concept in this thesis. There are different approaches to how we can sense the presence of other people in mediated environments and how we can evaluate this feeling of presence. Biocca, Burgoon, Harms and M. Stoner (2001) describe presence as the sense of *being there in other places* and *being together with other people*. They identify two categories:

   a) Telepresence, the phenomenal sense of 'being there' and mental models of mediated spaces that create an illusion;
   b) Social presence, the sense of 'being together with another' and mental models of other intelligence (i.e. people, animals, agents, gods, etc) that help us simulate 'other minds'.

There is extensive research in 'telepresence' in the field of virtual reality, which concentrates on fidelity to real-world appearance, video tunnels, tele-operators (robot arms) 'being there' or representations such as avatars to convey a sense of realism (Lombard and Ditton, 1997). The second perspective of social presence, however, associates the concept with a 'mental state', rather than an illusion of reality. Different aspects to the sense of social presence can be identified, such as mutual awareness, psychological involvement, behavioural engagement and cognitive states (Biocca et al, 2001). Presence can be sensed in non visual, but text-based virtual environments such as MUDs, MOOs, IRC chat etc. It also occurs when we read an interesting fiction book; it's the feeling of getting lost or wrapped up in the representations of the text – of being involved, absorbed, engaged, or engrossed in or by them (Lombard, 2000a). Through this process one can experience a 'willing suspension of disbelief', which can be described as the 'attitude by which the reader brackets out the knowledge that the fictional world is the product of language, in order to imagine it as an autonomous reality populated by solid objects and embodied individuals' (Ryan, 1999). Alternatively, this process can be seen as a 'willing construction of disbelief' (Gerrig, 1993; Gerrig and Pillow 1998); so as to emphasise the reader's act to assign value to a mental representation, as well as to subsequently reject it, if the representation contrasts their knowledge or beliefs about the represented world. Research in MUDs (Jacobson, 2002) has shown that the sense of presence is undermined when a virtual world resembles an existing one (e.g. a classroom), while fictional worlds enhance the sense of presence. However, the opposite happens with interpersonal communication; knowing people offline creates a remarkably greater sense of presence in online communication. In text-based virtual environments, interaction with others, rather than spatial representation, appears to be the significant factor in generating a sense of presence, i.e. *being with* rather than *being there* (Towel and Towel, 1997).

Our belief is that a sense of presence can also be achieved by simply knowing, *being aware* of other people's existence. I am 'online' or 'offline' as 'I am there' or 'not'. This is 'pure presence', a mental state of *being connected* or *in touch* with other people that can be communicated in symbolic and often abstract ways. What is interesting about this entirely 'mental' state is that you don't need to interact with people through technology, you just know that they are there for you to contact and that is already valuable: you get a sense that you are not alone. As an indication of the benefits of merely symbolic, mental presence, Nardi, Whittaker and Bradner (2000) report that people in their study of Instant Messaging use in the workplace found value in simply knowing who else was 'around' as they checked their buddy list, without necessarily wanting to interact with buddies.

Rettie (2003) proposes the concept of 'connectedness' as related, but not equivalent to social presence. The experience of connectedness entails psychological involvement: if I am aware that others are online, I feel like I am in touch with them even if there is no message exchange. Similarly, the exchange of 'goodnight' text messages creates connectedness (Rettie, 2003). Ijsselstein (2003) suggest that connectedness includes affective benefits such as stronger group attraction, a feeling of staying in touch, a sense of sharing, belonging and intimacy. Considering distance learning in particular, we know from the work of Whitelock et al (2000) that the presence of peer-group members can enhance the emotional well-being of isolated learners and improve problem solving performance and learning. Rheingold's discussion of Smart Mobs (Rheingold, 2002) highlights the overwhelming power of social cohesiveness that can be brought about by knowledge of the presence and location of others in both real and virtual spaces. According to Christiansen and Maglaughlin (2003) group awareness enhances the feeling of belonging to a group. Non-verbal communications, spontaneous interactions, informal and physical presence are all elements of face-to-face interaction that can promote a sense of community. Our aim is to capitalise on the positive social aspects of presence in a way that enhances the sense of being part of a group of people: a feeling of 'group belongingness'.

With the rapid developments in mobile and ubiquitous computing communication technologies presence has become a richer concept, incorporating attributes of a person's location or proximity to another person or location. Presence has also evolved from a simple online/offline (being there or not) description to include a variety of aspects, such as availability, location, communication preferences, device capability (Chakraborty, 2002), as well as more abstract 'states', like a person's intention and interest (Emilsson, 2001). In computer mediated communication (CMC) being aware of others' existence, plans, motivations, intentions, and state of attention is crucial. These entirely mental states embody a sense of rich presence. In effect, if I know that you're paying attention to me, and you know that I know this, we have a solid basis for communication; of course, this is simple to achieve face-to-face, but at a distance other, symbolic ways are needed to induce an analogous awareness. In the networked world, a sense of presence of colleagues or friends is facilitated by various communication tools, Instant Messaging (IM) being very commonly used. Presence is defined by a user status that answers the questions of Who (user), Where (location and device), When (preference and willingness), How, (device capability) and Why (information exchange, leisure, keeping in touch etc) (Chakraborty, 2002). IM is just one of many possible presence-based applications. For example, presence-enabled applications can facilitate safety-tracking of children by mobile phone. Other possibilities include location-based matchmaking and dating services, multiplayer games, and anything involving the collaboration of individuals separated in space and time.

Essentially, what presence enabled applications like IM do is to communicate a user's context. In figure 2.1, presence is defined by contextual parameters, such as activity, location and availability. This diagram outlines some presence functions based on an analysis of five Instant Messaging (IM) applications (ICQ, Yahoo, Odigo, MSN, Jabber) both desktop-based and wireless (Vogiazou, 2002). Context is closely related to time, location and the device that is used, as well as the user's availability, state of mind and identity. The list of contextual parameters can be extended and more attributes can be added to the diagram, which serves as an example, rather than a complete description of presence functions in IM. Availability is communicated in variable, yet very similar ways in different IM systems, defining the extent to which a person is available or unavailable (e.g. 'very busy', 'completely unavailable', or 'on the phone', 'away' etc) as well as communication preferences (e.g. custom messages: 'urgent communication only', 'available on mobile phone only', 'in meeting but can receive message' etc). In fact all these presence attributes are closely related to each other and often overlap. For example, the different types of being busy, such as 'occupied, urgent messages only', 'do not disturb' define availability, but they also reflect the user's state of mind. Some IM applications (e.g. Odigo) also display a person's mood (e.g. happy, bored, stressed, chatty, flirtatious etc) and intentions (e.g meet new friends, romance, play games, small talk, professional etc). A user can participate in many different work-related, project-related or self-constructed groups, which define his/her identity. In the same way that there are individual identities and user profiles, there can be group identities and profiles. The Faceted Id/entity project by the MIT Media Lab Sociable Media group addresses the need to personalise identities and communicate a different identity, depending on which group (or subculture in this case) a user is participating in (Boyd, 2001).

Context is a debatable concept in ubiquitous computing research because it can have several different meanings and definitions, often overstressing the importance of technology and underestimating the role of social context. One broad definition of context is: *Any information that can be used to characterise the situation of entities (i.e., whether a person, place or object) that are considered relevant to the interaction between a user and an application, including the user and the application themselves. Context is typically the location, identity, and state of people, groups, and computational and physical objects* (Dey, Abowd and Salber, 2001). According to Dourish (2001) there are two strands of context-aware computing within HCI research: a) physical based interaction and augmented environments and b) attempts to develop interactive systems around understandings of the social processes surrounding everyday interaction. Dourish argued that the second area, as the broad set of investigations into the relation between social interactions and technology, is an important form of context-aware computing that goes beyond the primary technological concerns and helps people to interpret and understand patterns of activity.

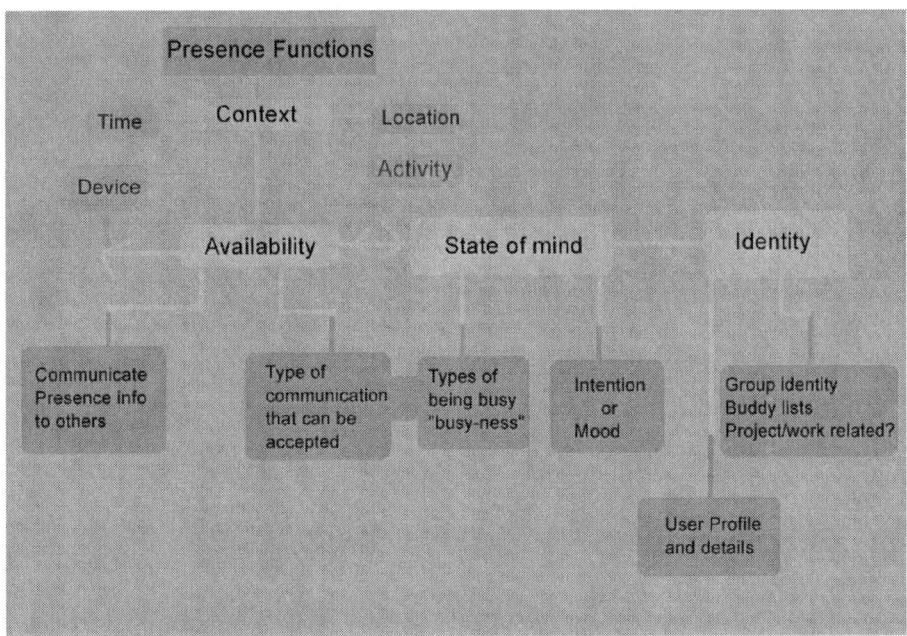

Figure 2.1 A set of presence attributes usually found in IM applications

Towards this goal, a couple of well established social computing research groups, as described below, have explored ways to communicate social cues and contextual information in a computer mediated environment in non explicit ways. They employ abstract representations of presence information, aiming to reveal some aspects of social behaviour in online environments.

Tomas Erickson and the Social Computing Group at IBM have designed 'socially translucent systems', systems that make perceptually-based social cues visible to their users, by supporting mutual awareness and accountability (Erickson, Halverson, Kellog, Laff and Wolf, 2002). 'Translucence', in contrast to 'transparency' indicates that the aim is not to make all socially salient information visible. It also stands for the notion that, in the physical world, cues are differentially propagated through space (Erickson et al, 2002). The Social Computing Group has developed several prototypes to illustrate the idea, for example the Babble System, a group discussion tool. A minimalist visualisation of people and their activities, what is called a *social proxy* indicates the level of attention. When people are either talking (type) or listening (click & scroll), their dots move to the inner periphery of the circle and then gradually drift back. This scales up with several conversation circles, as exemplified in the 'Landscape Proxy', where different categories of discussions are rendered as circles within a larger conversational space and they grow with user activity.

12   Chapter 2

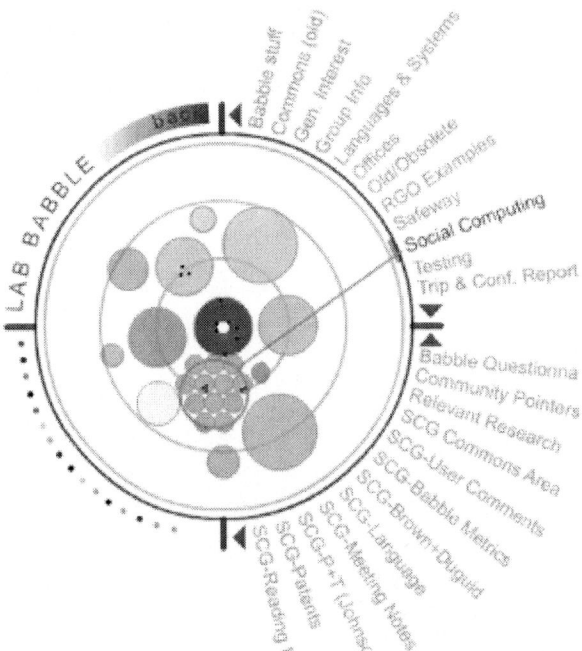

Figure 2.2 Screenshot from the Landscape Proxy of the Social Computing Group at IBM

The Sociable Media Group, at the MIT Media Lab has explored the potential of visual languages to convey social meaning. In one of their prototypes for visualising conversation, 'Chat Circles', presence and activity are made manifest by changes in colour and form (Viegas and Donath, 1999). Users are represented in space as a coloured dot with a name. Each person has a 'hearing range' that allows them to engage in conversation only with people in their vicinity, maintaining however a sense of a broader social environment and its activity. The users outside the 'hearing range' appear as outlined circles rather than fully coloured and their messages are not displayed. Users who have been idle appear as faded dots. The design of 'Chat Circles' indicates not only the number of people, but also their level of presence and participation with a more rhythmic and organic feel of the interface. Circles grow with text and slowly shrink and fade to their original dot size, as in real life conversations, where the focus is on the words said by the person who spoke last and, progressively, those words dissipate in the midst of the evolving conversation (Viegas and Donath, 1999).

Chatscape, also from the Sociable Media Group, uses visual metaphors that can be recognisable, such as roughness (e.g. a spiky shape) and smoothness (e.g. a round circle) to communicate different moods and meanings, like chaos and tranquillity respectively (Lee, 2001). In Chatscape, other users can modify a person's social identity profile by assigning attributes and in this way change the person's shape appearance, thus creating a 'reputation system'. This brings interesting social behaviour into the conversation.

Another, more recent visualization is 'The World as a Blog', a website which combines real time blogging activity with the blogger's actual location represented as a dot on the planet (Maron, 2003).

In light of the continuing research in ubiquitous computing, the notion of a user's context is still very slippery. It is very hard to deploy successful ubiquitous computing systems for predefined contexts of use, when these contexts are constantly changing and being defined by the users themselves. In Paul Dourish's words (2004) the *focus of the design is not simply 'how can people get their work done,' but 'how can people create their own meanings and uses for the system in use'; and in turn, this suggests an open approach in which users are active participants in the emergence of ways of working.* Although establishing such design guidelines for context is not a primary concern here, this thesis partly addresses the challenge by focusing on the emergent properties of design, what we identify as *design for emergence* in chapter 3.

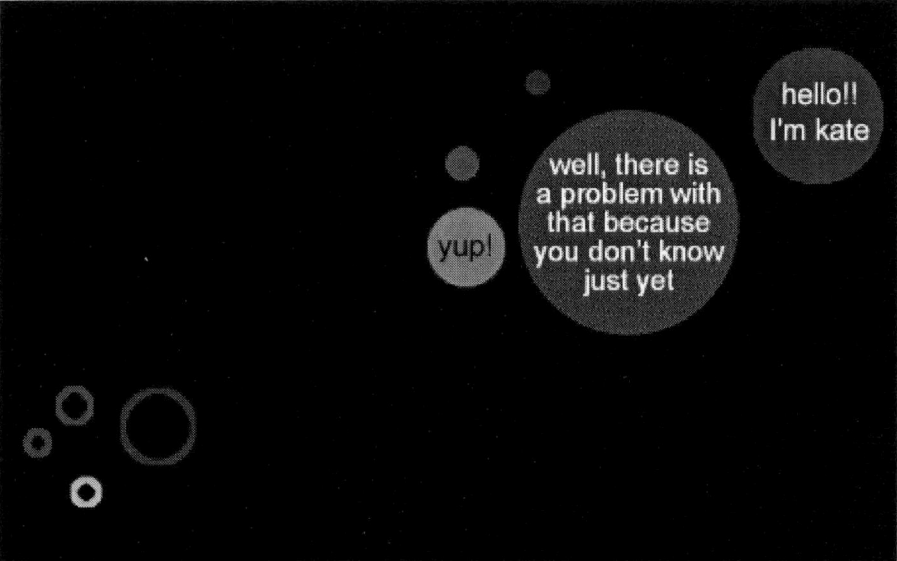

Figure 2.3 Screenshot of Chat Circles (from the Sociable Media Group http://smg.media.mit.edu/)

The aforementioned examples of social computing applications have been designed to be used by small groups. One of the early questions that puzzled us was: how can we sense the synchronous presence of many people? How can we achieve the feel-good factor of, for example, a concert, an event with large participation online? One of our initial motivations was to find out whether there is such a thing as a 'virtual crowd' and how the simultaneous presence of many other people can be a meaningful social experience.

A prime area where communities of thousands of active participants are already in place and use communication technology extensively is that of massively multiplayer online games and virtual worlds, such as Everquest and Asheron's Call (Microsoft, 2001). These are large groups of individuals creating a community with shared interests and real-time interactions. The sense of others being present is essential to simulate the social interaction within virtual worlds. There is a lot to learn from game design that can be applied to other contexts, such as the design of collaborative systems. In the next section we look into different forms of play in detail and explain why play is particularly interesting for our research.

2.1.2 Why play?

There is currently a clear trend in Human Computer Interaction research to study aspects of playfulness and enjoyment in interactive systems design, which is indicated by the publication of books like *Funology* (Blythe et al, 2004) and *Emotional Design* (Norman, 2004). Traditional usability approaches are now considered limited in scope by this new strand, mainly focusing on ease of learning, ease of use and task fit. A shift of focus towards the design of interactive products purposed not only for work, but for home and a whole range of everyday activities, has drawn attention to the user experience and aspects of engagement. While the buzzword 'experience design' has as yet no solid underlying theoretical and conceptual foundations, it nevertheless indicates a growing interest in motivation, pleasure and the play element of interactive products.

This thesis investigates play in relation to social interaction and group behaviour. Before outlining the reasons we focused on play, we draw a distinction between free, spontaneous play and games. In some languages, the words play and games are the same, whereas in English there is a distinction between the two. When attempting to draw the difference between play and games though, there appears to be more than one way to distinguish them and definitions vary a lot based on context. Salen and Zimmerman (2004) suggest that in one sense, 'play' is a larger term that includes 'game' as a subset and in another, the reverse is also true: games can be thought of containing play. In order to define 'game', the authors compare eight different definitions of 'game', by Parlett (1992, 1999), Abt (1970), Huizinga (1955), Caillois (1962), Suits (1990), Crawford (1997), Costikyan (1994) and finally Avedon and Sutton Smith (1971). Salen and Zimmerman noticed that all these definitions vary and the majority agrees only in that games have rules, which provide a structure out of which play emerges, and a goal, or some kind of final outcome. Their definition (2004) combines aspects of these approaches:

A game is a system, in which players engage in an artificial conflict, defined by rules, that results in a quantifiable outcome.

Their definition of play is much broader and abstract, trying to encompass game play, ludic activities and the sense of being playful altogether: *Play is free movement within a more rigid structure.*

Salen and Zimmerman propose the concept of transformative play, which is particularly relevant and interesting in the context of this thesis, because this kind of play *can overflow and overwhelm the more rigid structure in which it is taking place, generating emergent, unpredictable results.*

Playground games are great examples of transformative play because they involve a lot of improvisation and play differs every time. School playgrounds illustrate that highly engaging playful interaction is possible without the formally defined rules and goals of structured games. As Opie and Opie (1969) mention in their book on playground games, play is unrestricted, while games have rules; in the playground there is no need for an umpire, little significance is attached to who wins or loses and it doesn't even seem to matter if a game is not finished. School playgrounds are indeed a great arena for spontaneous play which is fun in its own right. These are spaces in which children develop their personal, communication and social skills through transformative play. The designs we describe later in this thesis aim to foster emergent, transformative play, not only in the context of the structure of play itself, but also in the context of players' behaviours and collaborative practices which occur within and beyond the game context.

Anthropologist Caillois (1962) distinguishes between formally structured games and free play with his concepts of *ludus* and *paida* respectively. *Ludus* represents rule-bound, goal-oriented play, while *paida* refers to spontaneous, improvisational play. These two concepts provide a framework in which we can position games based on their structure. As play edges closer to the *ludus* end of spectrum, the rules become tighter and more influential. Located on the other end of the spectrum, *paida*-based play eschews rigid formal structures in exchange for more freewheeling play (Salen and Zimmerman, 2004).

The thesis focuses on loosely structured games, which have the minimum rules possible, aiming to promote transformative play and to facilitate the emergence of collaborative player behaviour. Undoubtedly there is a growing body of research in games and play, primarily because even though 'play' is not considered 'serious', researchers have come to realise that there is a lot to be learnt from it and that play has certain significant values:

a) First of all play is important for its own sake, it's a fun, motivating and enjoyable activity.

Research has been carried out into what makes games, for instance, so motivating to play and how we can apply similar principles in other fields like educational technology. Malone (1980) has provided a general framework, a set of principles for the design of *intrinsically motivating* games (i.e. activities rewarding for their own sake rather than for the sake of some external reward). Malone identified the following values that make games fun: *challenge, fantasy, curiosity* and an appropriate amount of *informative feedback*. Also, feedback should be surprising sometimes to increase the curiosity of the player. They should be neither too complicated nor too simple with respect to the player's existing knowledge (Malone, 1982). The notion of challenge can be summarised in the provision of good *goal*, a goal with *variable difficulty* that is obvious and compelling at the same time. Fantasy is also seen as a means of escape from everyday reality. Malone believes that games are potentially superior to the traditional means of escape (movies, books, music) because they are participatory. Fantasy fulfillment frequently takes the form of symbolic exploration; discovering a big world, full of exciting things (Crawford, 1997). Malone identifies *curiosity* as the motivation to learn. He distinguishes between sensory and cognitive curiosity. While the former concerns the audiovisual aspect of a game and the representational system that is being used, the latter is more about a player's motivation to complete their knowledge structures of the game (Malone, 1980). Curiosity can be thought of as a drive to bring 'good form' to knowledge structures (Malone, 1982). Another, more design-oriented way of seeing this difference is distinguishing between the game interface (cognitive) and the game mechanics (sensory). The game mechanics are the physics of the world, which set the constraints on how we perceive our ability to navigate and manipulate the game's world and they affect cognitive processes (Jinwoo, Dongseong and Hoyoung, 1999) The game interface has more to do with player's perception of the game world. Games can evoke curiosity by providing an optimal level of *informational complexity*. They should be neither too complicated nor too simple with respect to the player's existing knowledge (Malone, 1982).

b) Play is a fundamental aspect of human nature, repurposing new technologies to unpredictable directions.

It is during play that we make use of learning devices, treat toys, people, and objects in novel ways, experiment with new skills, and adopt different social roles (Newman, 1990). While play has an indispensable part in children's cognitive and social development, it continues to accompany us in our adult life in a different form. Various types of games (e.g. board games, sports, computer games) illustrate the importance of play in adult life. Sometimes the distinction between those 'play times' and the rest of our life gets blurred. Play is an essential aspect of human communication for example, and this is evident with the great popularity of applications like IM, chat and SMS messaging. The success of SMS messaging, used to a large extent for playful communication, has been very influential for the design of mobile phones as we know them today: data services were introduced on top of voice and advanced messaging (EMS, MMS) facilities were developed. The mobile phone has become primarily a social device rather than a tool for work. Blogging is another example. Blogs are almost always personal, imbued with the temperament of their writers. The technologies to support blogging have been in place since the dawn of the web, yet it has not been until recently that this technique has self-organised itself into a playful social pursuit (Paulos, 2003). Humans engage fundamentally in social play, and these unpredictable uses of new media influence technological innovation.

c) As inherently social, play can strengthen bonds within a community and enhance learning and collaboration.

Play has always been inherently social, even before the advent of mobile phones, computers or any communication technology, as we see from school playgrounds. Before the emergence of video games, people got together to play Bridge or Poker or Monopoly or Dungeons & Dragons, to chat and socialize and have a fun activity to do together. Even games that require concentration and discourage table talk – like Chess or board wargames – were social, because you'd invite a friend to play with you, and the game was an excuse to get together (Costikyan, 1998). The internet has been the best medium for many-to-many communication so far. There are always people available online to interact and play with if one's friends are not. Several aspects in online multiplayer games promote socialisation, such as *diplomacy, social structure, communication, virtual community* etc (Costikyan, 1998).

The strong social aspect in play is important for collaboration and learning. We know from the deployment of distance education tools for Open University students (Scott & Eisenstadt, 2000) that it is not only possible, but desirable to foster relationships between isolated students by providing recreational social activities, such as a virtual Pub Quiz. We also know (Desouza, 2003) that workplace game rooms enhance tacit knowledge transfer at work.

d) The challenge of large scale collective play.

Massively multiplayer role playing games (for example Ultima Online, Asheron's Call) are an interesting field for analytical research because of their internal social and economic structures. The virtual world of Everquest has penetrated the economic activity of the real world as players trade game elements for real money on auction sites like eBay. Edward Castronova, of the economics department at California State University at Fullerton, studied thousands of EverQuest transactions performed through eBay to determine the real-world economic value generated by the 'inhabitants' of Norrath, Everquest's virtual world. Castronova (2001) discovered that Norrath's gross national product per-capita is $2,266. If Norrath were a country, it would be the 77th wealthiest country in the world, just behind Russia (Knight, 2002). This is indicative of how powerful and complex games can be, blurring further the boundaries between the illusive activity of play and everyday life.

Massively multiplayer gaming contains the force and influence that groups of people bring to real life and can thus have a social impact on people more powerful than real life can provide (Baron, 1999). However, the genre of massively multiplayer games is still under definition as game designers are looking into ways of building more interesting systems promoting different types of play, constructive and social interaction, without however yet having a clear image on how this could or should be achieved (Kosak, 2002).

Massive gaming communities can be great model for real life problem solving, uniting the intelligence and motivation of individuals in a powerful collaborative force. A striking example illustrating the potential of large-scale collective problem solving is the case of Cloudmakers, a group of more than 7,000 online puzzle solvers who proudly identified themselves in member profiles, home pages and email signatures as 'a collective intelligence unparalleled in entertainment history' (McGonigal, 2003). By forming teams, sharing diverse skills and knowledge and through intense collaboration, the Cloudmakers not only managed to solve puzzles that were designed for continuous game play of three months in just one day, but even attempted to solve real world challenges, thereby further blurring the distinction between virtual and real communities. For many of these players it was precisely this emergent collective intelligence rather than the game itself or puzzle solving that was immersive and highly motivating.

The above example is a great indication of possible extensions of play at large-scale to other arenas, such as collaboration and problem-solving. This thesis adopts collaborative social play as a testbed for emergent group behaviours. Before starting on the design board, we looked into research in social psychology to understand how spontaneous collective behaviours can emerge among small groups as well as large crowds. We outline relevant theories next.

2.1.3   Crowd behaviour and group dynamics

The crowd is a sort of medium if by that word one means the means for gathering and transforming elements, objects, people and things. As a medium, the crowd is also the site for the generation of expectations and the circulation of messages. It is in this sense that we might also think of the crowd not merely as an effect of technological devices, but as a kind of technology itself.

> Rafael (2003) on the SMS-linked crowd that assembled
> in Manila, Philippines to protest on January 20, 2001.

This work has been inspired from very early stages by emergent real-life crowd behaviours like the stadium phenomenon known as the 'Mexican Wave'(Eisenstadt, 2000). Every individual performs a remarkably simple behaviour (stand up, wave arms, sit down) in coordination with the person sitting next to them and without any particular goal, but just for fun, creating interesting large scale patterns in the stadium.

A critical mass of about thirty people is required to get the wave underway; then it subsequently expands through the entire crowd as it acquires a stable, near- linear shape (Farkas, Helbing and Vicsek, 2002) The Mexican Wave phenomenon is more likely to occur when spectators are not already over-excited, such as during flat periods in the game, and it works better in big crowds. For a scientist, the interesting specific feature of this spectacular phenomenon is that it represents perhaps the simplest spontaneous and reproducible behaviour of a huge crowd with a surprisingly high degree of coherence and level of cooperation (Farkas et al, 2002). We have been fascinated by the possibility of reproducing some form of Mexican Wave like behaviour online, with large numbers of real people performing a collective virtual activity together without any higher level coordination.

In public space events where one can be part of a crowd or a group of people, like concerts and festive celebrations, there is often a special atmosphere which can positively affect individual behaviour and feelings. On the other hand, in crises or challenging situations, people's actions and competitive behaviour are also influenced by the behaviour of others surrounding them. A striking example of this is the way people in a crowd can panic, e.g. when rushing towards a narrow exit during a fire, thereby blocking it. A classic experimental study of non-adaptive/self-defeating group behaviour was undertaken by Mintz (1951). Mintz's study showed that people change their behaviour according to their expectations of the behaviour of others, in relation to what actually happens in the process of a challenging situation. Mintz's experiment was valuable because it showed that non-cooperative behaviour in panics is not a result of violent emotional excitement as suggested by social psychologists in early crowd behaviour theories. Instead he explains 'the non-adaptive character of such behaviour in terms of people's perception of the situation and their expectation of what is likely to happen' (Mintz, 1951).

Figure 2.4 Alexander Mintz's experiment in 1951

Figure 2.4 shows how the experiment was carried out: each participant had to take their paper cone out of the bottle before it would get wet. Only one cone could come out of the bottleneck at a time and there was a reward or punishment structure with very little money.

His experimental study in the form of a game can be applied to a variety of disciplines. Mintz pointed out that if cooperative behaviour is required and a minority ceases to cooperate, then the whole cooperative pattern breaks in a vicious circle of results, because the needs of the individual conflict with the group strategy. This is useful to consider when designing games or other social activities requiring some kind of collaboration with possible tensions between the interests of the individual and the group.

Intergroup competition is another form of collaborative (or non collaborative) behaviour, frequently encountered in games, where groups of people act in certain ways and organise themselves in order to win other groups. But what does it mean to be part of a group, in social psychology terms? Research has proved that people change their behaviour accordingly if assigned as members of a particular group, even if the group identity is minimal, for example, based on random division. In the context of Social Identity Theory, Tajfel (Tajfel, 1970) performed several minimal group studies, where he discovered that group members acted in ways supportive of their, even minimal, group identity. His experiment fulfilled the following criteria, identified in 1992 by Schiffman and Wicklund:

1. No face to face interaction
2. Unknown personal identity of every group member
3. No particular advantage of belonging to one group or the other
4. No advantage or gain for the individual as a result of a particular position/action

Billig and Tajfel (1973) found that even when group members knew that group membership had been decided randomly (e.g. by tossing a coin) the results were still the same, i.e. supportive of the minimal group. These experiments are quite simplistic in relation to real life situations, but nevertheless indicate a tendency to identify ourselves in terms of 'we' when there is already some kind of social categorisation.

Taking a step back to the concept of the 'crowd' there is another theory that has investigated the effects of being part of a crowd, anonymity, issues of identity and personal responsibility in large group situations – the theory of deindividuation. Deindividuation is defined as the loss of self-awareness and evaluation apprehension in situations that encourage anonymity. Several studies have confirmed less acceptable social behaviour occurring when personal identity is hidden. For example, Zimbardo's (1970) 'electric shock' experiment with female N.Y.U. students giving 'shocks' (false ones), while either wearing a large name tag, or white hoods and capes, revealed that women with concealed identity pressed 'shock' buttons for twice the amount of time compared to women who were wearing name tags. Studies in other fields have reported findings consistent with Zimbardo's theory. Watson (1973) in an archival study of ethnographic records, found a clear correlation between cultures which indulged in highly aggressive practices towards their enemies and those which also regularly changed their appearance before battle in a ritual way (face, body painting or wearing masks). Other findings, from Diener's 'Trick or Treat' experiment (Diener, 1979), for instance, have provided more evidence for the theory of deindividuation. In this experiment, children would take more than one sweet when the experimenter was not present, even though they were prompted to take just one. When the experimenter was present or when there was a mirror just in front of the sweets they tended to be more obedient.

The focus on the negative effects of deindividuation has proved to be one-sided. The circumstances which are alleged to cause deindividuation may give rise to other forms of behaviour apart from aggression (Brown, 1988). In another experiment, Diener (1979) showed that a prior experience of activities designed to create a group cohesion (e.g. adoption of a group name, singing and dancing), subsequently led individuals to engage in more unusual and inhibited behaviours (e.g. playing with mud) than those who had an initial experience which made them feel rather self-aware. To convince on the contradictory possible effects of deindividuation, Johnson and Downing (1979) replicated Zimbardo's experiment, but had also women wear nurses' uniforms instead of hood and cape. Those wearing uniforms were less aggressive in 'giving shocks' than those not wearing uniforms. This also raises some interesting questions on how different fictional identities can affect behaviour in group situations. Diener (1980) suggested that factors present in some crowd situations – like anonymity, enhanced arousal, cohesion – lead people to direct their attention outwards and correspondingly less on themselves. In this way, people's behaviour becomes less self-regulated. The importance of these findings is that they show that being in a group, even in a primitive one, like the crowd, does not necessary lead to negative or aggressive behaviours as early crowd psychology had suggested. Rather people in deindividuated states, i.e. less self-aware, are more responsive to external, situational cues of how to behave than self-aware persons (Frank and Gilovich, 1988). Taking a further step, Reicher (1984) has suggested that crowd behaviour involves a change rather than a loss of identity. People might lose some sense of their personal identity, but their social identity sense, as members of a particular group increases. He also emphasised that crowd behaviour is very often an intergroup behaviour (e.g. rioters against policemen).

Recent research into the effects of computer-mediated communication (CMC) has also focused on anonymity and deindividuation. The Social Identity Model of Deindividuation Effects (SIDE) by Lea, Spears and Groot (2001) suggests that visual anonymity reduces the communication of interpersonal cues within a group, allowing certain social group information to become more salient. This has the effect of shifting perceptions of self and others from the personal to the group level, thus encouraging behaviour that is normative for the salient group. The self tends to be perceived and presented more in terms of similarity to a social group. In other words, depersonalised perceptions of self and others increase attraction toward group members and this process is stimulated by the dearth of individuating cues in visually anonymous interactions. This suggestion is contrary to early deindividuation theory, which associates negative, aggressive behaviours with anonymity effects.

It is precisely these positive and pro-social associations with a group identity, suggested by the SIDE model that we aim to promote through collaborative social play. We have used the informal term, 'group belongingness' in this thesis, to include all the positive feelings emerging when an individual associates with a group identity, when he or she feels part of a group of people, even without being able to see them or identify them physically.

We look at the balance between cooperation and competition as an important trade-off that we need take into account in our game design. The majority of commercial games have always focused on competition, in order to satisfy the need for self-projection and a sense of achievement against other players or an artificial game opponent, such as computer software. We have evidence of successful cooperative gameplay in simulations, where people need to cooperate to achieve particular goals, as in the Sims game (2002). In our games we aimed to have cooperative play as a challenge and we considered various scenarios as described in chapters 4-7 attempting to achieve a balance between having a competition element to enhance the challenge and enforcing collaboration as essential for succeeding in the game.

The social psychology theories briefly outlined in this section provide useful food for thought as well as inspiration for the design of social, presence based multiplayer games. Donath (1996) observed that the physical power of a crowd, both as an anonymous force capable of immense and destructive feats and as a force upon the individual in the crowd, is a salient feature that is absent in the virtual world. The quote on p.24 about the power of the crowd in the context of SMS-directed political protests suggests that there is a new space where large, 'real' crowds, empowered by wireless communication technology, can come into play, with unexpected emergent behaviours and consequences. Large scale cooperative location-based, or in other words mixed reality (both physical and virtual) play is yet unexplored although there is a significant amount of research in the area of mixed reality games in general, as discussed further in 2.2. Inspired by the aforementioned phenomenon of the 'Mexican Wave', we look at large scale participation as a key enabler and therefore an important starting point of our work. The following paragraph provides some insights in how self-organisation can occur and defines emergence.

2.1.4 Emergent self-organisation

In his book *Emergence,* Johnson (2001) refers to the emergence of ant colonies, flocks, swarms, city neighbourhoods and how scientists have eventually have come to understand such phenomena well enough to reproduce them through artificial intelligence and simulations. We also know that disciplines like biology (Holldobler and Wilson, 1994) and mathematics (Wolfram, 2002), have explored in depth complex collective self-organization and emergent bottom-up behaviours. The most primitive form of a 'society' in the broadest sense of the term is the anonymous flock (Lorenz, 1967). In many species which form large flocks the individuals never come nearer to each other than a certain minimum distance; there is always a constant space between two animals of the flock. According to (Johnson, 2001), large patterns and complex behaviours can emerge when multiple entities interact dynamically in multiple ways, following local rules and oblivious to any higher level instructions. Local interactions are a key term in understanding the power of swarm logic. Another important point is that '*more is different*'. In the case of ant colonies for example, there has to be a critical mass of ants roaming an area and communicating through pheromones for the colony to be able to asses its global state and to calculate the most efficient route to a food source.

There are four distinct facets of distributed being that supply 'swarm systems' (both biological and artificial) their character (Kelly, 1994): a) The absence of imposed centralized control b) The autonomous nature of subunits c) The high connectivity between the subunits and d) The webby nonlinear causality of peers influencing peers.

The relative strengths and dominance of each factor have not yet been examined systematically.

The most prominent characteristic of emergent phenomena is that the unexpected self-organisation resulting from many dynamic simple interactions is different and more important than each one of those interactions. According to John Holland (1998), emergence is above all a product of coupled, context-dependent interactions. Technically these interactions, and the resulting system, are nonlinear: The behaviour of the overall system cannot be obtained by summing the behaviours of its constituent parts. Holland makes an important point about context-dependent interactions: it is not the sum of individual parts but their interrelationships that make a difference and result in macrobehaviour in emergent systems.

While scientists have been successfully recreating and simulating swarms and flocks in artificial systems in the last 20 years, a new interesting challenge is to explore such behaviours among large numbers of people, empowered by communication technology. Rheingold has suggested in his book *Smart Mobs* that internetworked groups of humans can exhibit emergent prediction capabilities (Rheingold, 2002) and thus demonstrate self-organizing dynamics. The research work of this thesis aims to provide a foundation for further research in technology-mediated self-organisation in the real world. We outline some relevant examples next.

An interesting parallel between the activity of ants searching for food and online blogging activity has been drawn by an active 'blogger' (Hiler, 2002): adding links to interesting websites (blogging) reminds us of the activity of leaving pheromone trails to sources of food. While *ants* quickly find the closest food sources, and work together to consume it, *bloggers* quickly find the most interesting news stories, and work together to cover/analyze them. *Ants* focus on the closest food sources, consuming them until they are all gone. In a similar manner, *bloggers* tend to focus on the most interesting news story, covering them until there are no more angles or insights left. The relatively recent emergence of weblogs is indicative of the potential of self-organising dynamics among distributed, yet connected individuals, in accordance to all four of Kelly's factors as described above.

A remarkable example of simultaneous emergent behaviour, interestingly again in the context of play, is Carpenter's event at a conference of computer graphic experts in Las Vegas in 1991, described by Kelly in his book Out of Control (1994). Five thousands people were sitting in a huge conference room, each waving a cardboard wand with reflective material, red on one side and green on the other. The screen in the auditorium displayed a high contrast, real time video view of the audience and a computer counted the total number of the green and red wands and used this value to control software. Carpenter, a graphics wizard, instructed the people on one side to hold the wands with one colour and people on the other side the opposite colour. He then launched a game of Pong, only in this version the red side of the paddle moved the wand up and the green moved it down. Each wand was just a vote and each move of the paddle was the average of several thousand of player's intentions. The audience managed to self-organise and learn to synchronise better when Carpenter increased the speed of the ball. Likewise the audience managed to form numbers within a circle on screen and even fly a virtual airplane (this proved harder due to delays in immediate feedback). Kelly compares this behaviour to birds' flocking, only that conference attendees flocked self-consciously. Carpenter's experiment fulfils the fundamental characteristics of emergent behaviour:

1. *More is different*: 5,000 people interacting.
2. *Local interaction*: Participants observed what their neighbours did.
3. *No central control*: Carpenter only gave initial instructions, thus acted as facilitator rather than as organiser – it was completely in the responsibility of the audience to self-organise from that point.
4. *Simple rules*: turn the wand to green or red.
5. *A macrobehaviour emerges*: crowd synchronization controls a Pong paddle, the formation of a number and the flight of an airplane on screen.

However, what is significantly different between animal flocking behaviour and this example of self-conscious crowd synchronization is the fact that people in Carpenter's auditorium had the overview of the whole group of 5,000 on screen and it was precisely this feedback on screen that allowed them to coordinate. When feedback was not immediate enough (e.g. sometimes in the flying airplane exercise) the group mind was getting confused. In flocks however, each member responds to the reactions of their neighbours, there is no overview of the whole and the only guides are: don't bump into anyone, keep up with your neighbours and don't stray too far away.

The participatory experience of seeing 5,000 individual actions displayed on screen gives a sense of large scale presence. An overview of massively multiplayer games reveals that there is yet no game online in which a player can experience such a synchronous massive participation of other people and interact with them in some way. Although massively multiplayer games are played by tens of thousands of people simultaneously online, usually players interact in small numbers, for example five or so at a time. The player's view is limited by an artificial horizon in the radar visualization: this conveniently narrows the immediate scope of events requiring urgent attention, but restricts the total immersion effect that was achieved in Carpenter's event. Maps of those virtual worlds show more people, but nowhere can one find the sense of the actual crowd of thousands simultaneously in the game.

In the last four sections (2.1.1- 2.1.4) we tried to understand aspects of presence, play, social and self-organising crowd and group behaviour. The next paragraph identifies opportunities and challenges in creating playful social activities mediated by mobile and ubiquitous computing technologies. We draw on existing examples of wireless location-based games and social software and discuss why these applications that have recently come into play are appealing for research.

2.2 The challenge of mixed reality collective experiences

Recent studies (Paulos and Goodman 2004) have explored ways to represent and communicate our relationships with people we encounter regularly in our everyday lives, yet hardly ever get to interact with, our encounters with 'familiar strangers' (Milgram, 1977). These studies have illustrated that our perception of who else is around us is a significant factor for feelings of comfort and reassurance in public spaces. In this context, this thesis addresses possible social implications of mobile and ubiquitous computing technologies such as the feeling of being part of a group, based primarily on the sense of others being present in mixed reality (both physical and virtual) spaces. The research explores the boundaries between the physical and virtual world: what kind of engaging social experiences can emerge in the real world based on the awareness of individuals participating in a parallel virtual experience? Does virtual presence penetrate physical presence in any way? How can we design for the emergence of spontaneous social behaviour and group interaction in public spaces?

New opportunities have also emerged for individuals and groups to communicate and coordinate their activities spontaneously in urban environments. In this context, we have been exploring how spontaneous individual and group behaviours can emerge in the real world through playful collective experiences. For example, the recent Flash Mobs phenomenon (Kahn and Kellner, 2004) illustrated that people do not hesitate to perform certain acts in public together with many others, which otherwise would have been quite embarrassing. In fact, people participating in those events have appeared very engaged and amused. Observations from participating in two such acts of spontaneous play have informed this research and are described in the next paragraph 2.2.1.

The following sections cover an overview of wireless and location-based games and social software applications, drawing from both the industry and academia, in order to map and gain an understanding of the most prominent characteristics in this emerging and rapidly developing field of research.

2.2.1 Participant observation in Flash Mobs

Flash Mobs were organised through email lists and emerged as a trend in spring 2003, first in the US and then around the world. A flash mob is an event in which a crowd of hundreds appears without warning, performs a surreal act together and then disperses in the blink of an eye (Telegraph News, 2003). The trend began in New York in June 2003 with the first ever Flash Mob, when more than 100 people mobbed Macy's store in Manhattan. All requested a 'love rug' for a 'suburban commune', sending shop assistants into a confused panic. Flash Mobs exploded within two months and made the journey from underground to mainstream at fiber-optic speed (Felch, 2003). These acts of spontaneous performance follow specific instructions which are distributed in specified meeting places at the very last minute. Many people were very enthusiastic about these events:

> *It's interesting being part of something, getting attention and being a bit cutting edge. For me, it's also become a meeting place for new friends. I'm a pretty reserved, stand-at-the-back kind of bloke but I enjoy being there and seeing the expression on people's faces.*

(A Flash Mob participant in Telegraph News, 2003)

The spontaneous and bizarre nature of these events in public spaces was nothing new though. The Situationists (SI Archives, 2001), an avant-garde art movement and network of European artists and writers in the 1950s and 1960s, were famous for their surrealist street events and performances in public spaces. Also, numerous 'reclaim the streets' type of events have been organized by independent groups all around the world ever since, often involving unusual performances and spontaneous acts of play.

The author of this thesis participated in two Flash Mobs events in London in 2003. The first of these was very successful and engaging for the hundred and fifty or so participants. They performed a fun activity together: greeting passing trains and boats on top of the Embankment Bridge in London on 23/08/03. Most fascinating was everyone else's reaction to what was happening, including the train drivers and tourists on boats, who greeted the crowd back with excitement. The feeling of participation and social fun was very strong. The build up of anticipation prior the event was also important. In the arranged meeting location, which was communicated by a forwarded email, a lot of people were hanging around. This is not unusual for the particular spot, outside the Royal Festival Hall in London. It was hard to tell who was there for the Flash Mob or just enjoying the nice summer weather, some people, including the author, were looking around persistently trying to guess who was part of the Mob and who wasn't – this in itself was very exciting and it felt as if something was just about to turn over the world's normality.

The second Flash Mob in Covent Garden, however, was not as engaging because of the context: the market was so crowded anyway, that it was not very easy to distinguish the Flash Mob participants from the usual passers by when it all started. So the crowd did not have as strong presence in this case and the author felt more embarrassed to perform the instructions of the mob (to greet others in foreign languages while going around the square in circles).

Based on observations during the two Flash Mob events in London and news coverage of other such events around the globe, our conclusions on what makes participating in a Flash Mob an engaging social experience can be summarised as follows:

Figure 2.5 Flash Mob in London (Image from www.fahrbach-web.de/International/v2/whatmob.htm accessed on 23/08/03)

1) It is a pointless activity. This means that by participating what does not feel that he or she serves any political purpose or benefits any particular organisation or individual. It is all about being part of an activity that is fun in its own right, without the need to justify it. This freedom is reflected in the diverse audience that shows up for these events.

2) It involves synchronisation and simultaneous performance with many others, based on simple actions. Simple actions, such as clapping, waving, shouting, singing, dancing in a certain way etc can be easily performed simultaneously by many people together. It is this sense of participation that makes the activity engaging. However, not all activities are like that, for example in the London Flash Mobs the instructions also said to 'click your fingers every time you say or hear the letter y'. Such instructions were performed sparsely and were not as engaging, because they were too complex for people to follow at the same time. These activities became a subject for conversation among Flash Mob participants.

3) It is self-organised. Although instructions are given out, there are no 'leaders' in a real Flash Mob. As the organiser of the first event in New York said in Wired News (Delio, 2003): *The idea is mine, and I write the e-mails, but I don't think of myself as the leader of the mob. In my mind (the mob) is led by whoever forwards the email around. People make the mob through whoever they know.*

4) It is massive. Consider the difference between the two events in London: the Covent Garden one was not as successful because the mob did not have a strong presence in the already crowded square. A large number of people present is necessary to evoke a sense of participation and to decrease feelings of embarrassment and shyness.

5) The presence of non participants is important. This point is related to the one above: not only you need to have the presence of many people together, but you also need 'witnesses' of the event. Flash mob participants enjoyed seeing others' reactions and surprise and this increased their sense of participation, a distinction between 'us' and 'them', in other words, people in the know versus passers-by.

6) It is quick, fun, surreal and does not require too much commitment. There is no obligation, apart from being at a certain place on a certain time and spend about half an hour altogether to participate in these events. It is also an alternative form of entertainment, something unusual that breaks everyday life patterns in an enjoyable way and only for a short time, as the whole thing only lasts ten to fifteen minutes, so people do not get bored.

These characteristics also apply to the social and participatory large-scale play we are interested in. The most important aspect is to try and maintain people's interest in the experience: in the same way that flash mobs need to be innovative through a new idea every single time in order to be sustainable, any form of spontaneous group play also needs to be diverse, a different experience every time.

In order to achieve this, the experience must incorporate some element of *challenge*, like mystery and anticipation. For instance, Flash Mob participants were restless in anticipation outside the Royal Festival Hall in the second London Flash Mob, trying to guess who was part of the event and who was a passer-by by exchanging looks. So there is an interesting space to experiment with: who is part of a mixed reality experience should *not be obvious* until a later phase. Details about participants' physical presence should be *revealed gradually*. Another interesting aspect is the feeling of being part of a group, or what we have called 'group belongingness'. Flash Mobs could achieve a balance between *participating in a large crowd activity anonymously* and *being social, part of a small group of people*. Participants did weird things collectively in hundreds, but they also met other people and socialized with their friends, often heading to the pub after the event. A *critical mass* is necessary for a Flash Mob type to be successful: we found that the crowd gathering for the event needs to have a visible and strong impact on passers-by and people at the same location who are unaware of the Flash Mob. The crowd needs to outnumber non-participants by far, so that being in the larger group becomes an advantage. It is a kind of 'us versus them' division, where the feeling of participation becomes stronger as more and more people do the same funny act together.

When considering playful behaviour in public, we need to address the trade-off between the exciting feeling of *challenge* and the *embarrassment* of revealing something personal or confronting complete strangers. How much are we prepared to give up of our privacy in the shake of adventure? People need to have trust that they will not get hurt. What guarantees of *trust* did Flash Mobbers have? The fact that the organisation of these events was anonymous, without political connotations and that there were no commercial or other profit interests behind it, was sufficient for participants not to feel that they are being used for another purpose. The decline of Flash Mobs was partly brought by huge media attention, because this is when people felt that they are being exploited and they lost their trust in these initially innocent and fun events.

2.2.2 Wireless location-based multiplayer games

With the advances of wireless technology new types of games have emerged, often appealing to a broader and more casual and diverse audience than dedicated online gamers. Similarly, these technologies are becoming vehicles for social communication and the formation of social networks, based on location.

Gaming applications based on mobility open up new interesting opportunities, as people have some time to spare for play when waiting for the bus or commuting. Use of commute time partly explains the success of mobile games in Japan (McLorinan, July 2001). The usage patterns vary, but the most common pattern for mobile gaming appears to be frequent-but-brief interactions.

An early interdisciplinary EU funded research project on mobile play and social software in 1998, FLIRT (Flexible Information and Recreation for Mobile Users) explored the relationship between information flow and the urban environment (Raby, 2000). The project focused on awakening imagination, blurring the real with the virtual through fictional narratives. Interestingly, many of the playful ideas presented by FLIRT (figure 2.6) were transplanted within the following two years in the form of real dating services and location-based commercial games, like Botfighters.

In the 'Botfighters' game, created by the Swedish company 'It's Alive' (It'sAlive, 2002), players sent 'shoot' messages to each other depending on their proximity. The game combined online with mobile gaming. As a player, you would first build a robot on the website and then the actual game would take place in the street, using the mobile phone as radar to locate others and to send them various predefined 'attack' messages. When their mobile phones were on, players would receive SMS messages about the geographic distance of other players. Other early pioneering companies that developed location-based games were the Swedish BlueFactory (BlueFactory, 2000-2002) and the UK based Digital Bridges (Digital Bridges, 2001). Mitsubishi/Trium prototyped a real-time location based hide-and-seek game, Manhunt (Bruce, 2002). The Danish company

Unwired Factory (UnwiredFactory, 2001) developed a location-based treasure hunt game, the TreasureMachine, as well as another game called Zonemaster, where the aim was to conquer 'zones', parts of the city. The first graphically interesting mobile games emerged in Japan, developed for the i-mode platform. In the early Samurai Romanesque (Scuka, 2001), a massively multiplayer Java based game with reference to actual historical places, players practiced sword fighting and socialized. Although thousands could be playing the game, players interacted with no more than one other game character at a time. Samurai Romanesque integrated real-time weather data, provided by the Japan Weather Association, so that game settings would change as the real world weather changed; for instance, when it would really be raining, a character would move slowly, as the roads in the fictional game world would be muddy. In the early Japanese 'fishing' games on i-mode, commuters would pick up elements from their surroundings on their way to work and then get different scores.

Figure 2.6 Pixel kissing, an early concept of the FLIRT project (by Tony Dunne and Fiona Ruby from http://www.dunneandraby.co.uk/designing/FLIRT/FLIRT.html).

An interesting, more recent example is Mogi, a collecting game developed by the French company Newt Games and played in Tokyo. The game provides a data-layer over the city of Tokyo. As you move through the city, if you check a map on your mobile phone screen, you'll see nearby items you can pick up and nearby players you can meet or trade with. The game ties the desktop to the mobile internet. Hardcore players using web terminals can command mobile casual players to work in a team effort. In this way the web interface becomes a means for the hardcore players to orchestrate the experience for the mobile (casual) players. This kind of interdependence, with the right interface, allows for players with variable skill levels to play together (Hall, 2004). The game also affords for opportunistic interactions, as described in the following account of a regular player (Baron 2004b):

Over the past month, I bumped into a player who turned out to be the creator of the game, I had to race to pick up a flag that had been put on the map at equal distance between me and another player to encourage us to meet.

Conqwest (2005) is a team-based treasure hunt game, in which players try to claim areas of the city by bringing their totem-like large animal figure to that location. The interesting physical, yet very low tech aspect of the game is that they need to look for a certain type of black and white pattern displayed on stickers, flyers, billboards and cars in different sizes at different places in order to get the area 'code' or web address, necessary to claim that space. Some other recent examples of wireless and location based games are: Undercover, a massively multiplayer, persistent game for mobile phone users in Hong Kong, Gunslingers a multi-player mobile game where players move around, track and engage enemies within their vicinity, the Go Game, a team based adventure game and puzzle solving games Navigate the Streets and Mad Countdown, among many others (Baron, 2004a).

Another interesting domain is 'pervasive gaming', using a variety of media to engage players in an activity where the game, rather than the players themselves, controls the 'where', 'when' and 'how' of playing. Majestic by Electronic Arts (EA), for example attempted to blur the distinction of game fiction and everyday reality. Majestic however failed in its attempt to attract an audience of casual gamers and did not appeal to hard core gamers, partly because the player had little control on the time and circumstances of the gaming activity (Kushner, 2002). The similar Nokia Game (Nokia game, 2001) on the other hand, is a successful example because of its international and well designed advertisement, played by 600,000 people in November 2001. The Nokia game was free and it engaged players in solving a fictional narrative, which evolved in real-time during three weeks. More interesting than the game play itself was the intense communication among players all around the world and the really rapid spread of information through all sorts of facilities: chat, mailing lists, websites etc. Most recently, the Vienen Por Ellas (They come for them) pervasive game played in Chile, has also been successful. Users tried to win against 'aliens' by solving quizzes, answering questions, finding the clues, etc. The game was played via SMS, voice messages, Web sites, WAP, moblogs, MMS, ringtones, etc (Baron 2004a).

This list is far from being extensive and shows a great creative and commercial interest in an emerging area. There is significant research in the area of location-based multiplayer games in the academic world, which is growing as one of the strands within ubiquitous computing. A well known example, by a group of artists called Blast Theory and the University of Nottingham, is 'Can You See Me Now'. It is a game fusing location-based with online gaming, incorporating GPS positioning and a wireless network. Players are playing on their computers, but their avatars are being hunted in the streets by real people (Benford 2002). Another, more recent, Blast Theory project called 'Uncle Roy all around you' explores problem-solving between online and mobile players (Flintham, Anastasi, Benford, Hemmings, Crabtree, Greenhalgh, Rodden, Tandavanitj, Adams and Row-Farr, 2003) through a fictional mystery. A similar mixed reality game, using more heavyweight technology though, is NetAttack (2004) developed by the Fraunhofer Institute of Technology.

The Schminky project (Reid, Hyams, Shaw and Lipson, 2004), by the Mobile Bristol team at the Hewlett Packard Laboratories in Bristol, UK introduced a sound-based multiplayer game into the heart of a social space, a café. The Schminky user trials showed that the game helped to facilitate initial social contact among strangers and that players frequently drew together, even though they could play the game anywhere in the café. The ambiance of being in the café and around other people enriched the experience even in single-player mode.

CatchBob! (Nova and Girardin, 2004) is designed to facilitate collaboration among people working together on a mobile activity. It is a group hunt game based on location awareness. Participants can see the location of their partners with a colored dot on a map, displayed on an iPAQ pocket PC or tablet pc. They also see how close they are from the object they are trying to locate. What is particularly interesting about CatchBob!, is that it enables visual communication: if a participant taps a dot (that represents a person) with his/her stylus on the device, (s)he can draw a vector that corresponds to a suggested direction for his/her partner, like saying 'go in this direction' (Nova and Dillenbourg, 2004). The game also serves as a behavioural data collection (users' paths, messages) tool. CatchBob! illustrates an interesting solution for small group coordination via updated dots and suggested paths on a map.

In a relevant research arena, researchers at Glasgow University have developed 'Seamful Games' (Chalmers and Galani, 2004), to explore positive uses of infrastructure challenges. Their goal is to harness negative aspects of infrastructure technologies (such as GPS inaccuracy), which are normally concealed and unexplained, and present them as game features allowing users to explore and understand them. In the Seamful Game players travel around a designated area collecting digital 'coins' and upload them in exchange for points. They must develop an understanding of the network coverage and the effect of signal strength in order to successfully play the game. In this way the patchy network coverage, which is usually seen as a problem to be overcome, or worse ignored, is turned into a feature of the game (Chalmers, Bell, Brown, Hall, Sherwood and Tennent, 2004b). The author of this thesis tested this game at the UbiComp 2004 Conference and noticed a clever dimension: the game encourages players to interact with each other and collaborate. Players can upload coins simultaneously with other members of their team when they meet at the same location, to receive a points bonus (each player receives the cumulative total points for the upload) and the more team members participate, the more points they stand to gain.

2.2.3 Location-based social software

Social software is a term with variable definitions, extending the meaning of the earlier terms such as 'groupware' and 'collaborative software' to include all kinds of software that support group interaction, even if this interaction takes place offline (Allen, 2004). Clay Shirky organized a 'Social Software Summit' in November 2002 and since then the term has been broadly used. Shirky also wants social software to 'explicitly try to include online support for both lightweight social value (e.g. del.icio.us) and offline interaction (e.g. Dodgeball, PacManhattan) in the definition' (Allen, 2004).

The Meatball Wiki site (2005) lists several online community sites, which help the user to find others with similar interests and build up networks of acquaintances online, based on the basic principle of 'who knows whom' or in other words, 'who is a friend of my friend' and profile information (table 2.1).

Table 2.1 List of social software websites

Friendster.com	Friendfinder.com	(Meatball Wiki, June 2004)
Myspace.com	Tickle.com	
Tribe.net	Yeeyoo.com	
Hi5.com	Orkut.com	
Yafro.com		

Other social software uses focus on sharing and finding resources based on patterns of usage, for example photos (Flickr.com) and website bookmarks (e.g. del.icio.us).

Most interesting for our research, are the emergent mobile and location-based social software applications. Dodgeball (2005), for example, is a location-based messaging system which serves to notify users when their friends are in vicinity so that they can meet up. WhoAt (2004) is a wireless location-based dating and match making service. Location-based dating services have existed in Japan as early as 1998 with the huge success of the Lovegety keychain devices (Iwatani, 1998), which signal when another Lovegety owner of the opposite sex and a compatible profile is within fifteen feet. Owners can set the device to show display lights according to whether they are in the mood for a simple chat, ready to sing karaoke, or want to go all the way up to the 'Get2' mode, in which anything the couple wants goes (Reuters, 1998).

Another example of location-based social software, the UK based friends network Playtxt.net (2004), uses six-degree's-of-separation via mobile phone to meet friends, or friends of friends. Mamjam (2003) is another location-based messaging platform for mobile phones, which works with SMS. Socialight (2004) is a location-aware mobile social networking platform that allows people to connect with their friends and friends of friends. ImaHima (2004) is a community and instant messaging service allowing users to share their current personal status (location, activity, mood) publicly and privately with their buddies through mobile phones. When you select the 'update' link on your mobile's ImaHima menu, everyone on your buddy list knows, for example, that you are within a few blocks of Shibuya station and are free for lunch (Rheingold, 2002). A more business oriented application, BuZZone (2004) uses Bluetooth to find users with a profile matching a set of search criteria. Mobile communities are already in place with services such as UPOC (2005), a platform for users of mobile phones, Internet phones and text pagers in the US to send text and voicemail messages to groups of people at the same time. UPOC users form groups around shared interests, like musical performers ('Destiny's Child'), television shows (HBO's 'Sopranos'), information useful to people on the go ('NYC Subway Alerts') and gossip ('NYC Celeb Sightings'). The number of Mobile Social Software (MoSoSo) services is constantly growing as indicated by the list in a relevant blog entry (Meskill, 2004).

On the academic front, an early research project by the University of Oregon was Proem, a wearable system for profile-based cooperation that enabled users to publish and exchange personal profile information during physical encounters. The Proem system was used to initiate contact between individuals who had never met by identifying mutual interests or common friends (Kortuem, Segall and Thompson, 1999). WALID, by the same research group, was a grass roots community cooperation project, implementing a 'digitized version of the timeworn tradition of borrowing butter from your neighbour' (Kortuem, Schneider, Suruda, Fickas and Segall, 1999b). The aim was to assist people with a broad array of tasks, ranging from personal scheduling to task planning. Personal wearable agents were used to enable goal directed cooperation during physical encounters of people with selfish and conflicting goals, such that cooperation would lead to mutually beneficial results. Social Net (Terry, Mynatt, Ryall and Leigh, 2002) was another interest-matching application that used mobile devices outfitted with RF-communication to record the time and duration of encounters between users. The application searched for patterns of physical proximity between people, over time, to infer shared interests between users. When two users who did not know each other were assumed to share interests, Social Net would consult their lists of friends to seek a mutual friend between the pair. If one was found, the mutual friend would receive a suggestion to introduce the two. What distinguished Social Net from other academic and commercial projects of the kind was this attempt to balance the affordances and capabilities of the technology (i.e., the ability to detect long-term trends of collocation between people) with the natural ability for people to mediate interpersonal interactions (e.g., deciding if, when, and how two people should be introduced).

Most recent, Serendipity (2005), an MIT all-in-one application aims to instigate interaction between potential friends, partners and colleagues, based on user profiles and areas of knowledge. The Serendipity project further aims to research and model how information flows across a social network (Eagle, 2004).

Other applications focus on public authoring: the idea of personal or community mapping of urban space by posting messages on real locations where others can retrieve them, sharing experiences of the city and creating a collective portrait of it. One of the earliest applications, GeoNotes was developed by the HUMLE-Lab of the Swedish Institute of Computer Science. GeoNotes was a system on pocket PCs with which users could annotate physical locations with 'virtual notes'. These could then be accessed by other users in vicinity. Drawing from the concepts of posters, signs, notes and graffiti, the system allowed ordinary users to provide, update, remove and comment information in various places. In this way it created social awareness in physical space that encouraged play, expressiveness and personal identity formation (Espinoza, 2001). Urban Tapestries (2005) and PDPal (2003) are another two example projects of mediated public authoring. In Urban Tapestries in particular, experimental ethnographic methods are used for investigating the relationships between communication technologies, users and the socio-geographic territories around them (Silverstone and Sujon, 2004). Mudlondon (2004) is a kind of collaborative mapping project, based on a model of London, with grid location data about places and the connections between them. Users can connect new places to the model, augmenting it with their own mental map, annotating with descriptions, urls etc through an Instant Messaging interface. Another interesting project is the MobiTip service (Rudström Svensson, Cöster and Höök, 2004), developed by the Swedish Institute of Computer Science, a mobile collaborative filtering or recommendation system that makes use of relative positioning using Bluetooth. MobiTip allows its users to express their opinions and comment on anything of interest in a defined environment (a shopping mall). Comments given by one person are made available to another when users pass each other, when they approach connection hotspots, or on demand. Users may enter their own opinions, as well as inspect and react to tips from others. The presence of other MobiTip users, information hotspots and other Bluetooth devices appear and disappear on the user's device as they move around. MobiTip is the first collaborative filtering system that we are aware of at the time of writing that uses both similarity among user ratings and proximity to other users to determine what kind of information is displayed on the device.

All the above examples indicate research directions where the global meets the local and the virtual meets the physical:

The exploratory movements of locative media lead to a convergence of geographical and data space, reversing the trend towards digital content being viewed as placeless, only encountered in the amorphous and other space of the internet.

(Locative.net on Transcultural Mapping, 2004)

A hybrid application, bringing together local and global information about networked places and people is Plazes. It is a global location-aware interaction and geo-information system, in which users can 'discover' Plazes, where a Plaze is a physical location with a local network - private or public, wired or unwired. A Plaze constitutes of the information about the actual location like pictures, comments and mapping information, as well as the people currently online at that Plaze (Kellner and Petersen, 2004). Users upload this information when they are physically at the Plaze. If the networked place is registered for the first time in the system, then it is their 'own'. This brings to light important issues of ownership of information that such a collaborative annotation/mapping activity involves: anyone physically present at a location can incrementally complement or alter the information about it. But Plazes works on the principle of wikis (Remy, 2002), therefore if someone writes something bad or wrong, someone else will probably correct it. In the context of our research, what is mostly interesting is the integration of a real world activity (annotating a physical space) with the presence of other people (being able to see who else is online at the same location or elsewhere), which creates opportunities for social interaction and collective collaboration (or competition, for example for the 'ownership' of that particular Plaze).

A remarkable example of wiki-based collaboration dealing with a large-scale natural disaster is the website Scipionus (Mendez and Stoll, 2005). At the time of thesis writing, the website displayed maps (powered by Google Maps, 2004) of the area affected by the hurricane Katrina in the US with information about the state of the damage at specific locations. All of the information on the maps was provided by ordinary citizens, creating a giant visual 'wiki' page, attracting tens of thousands of visitors (Singel, 2005).

2.2.4 A categorization of ubiquitous social experiences

The following categorization considers a sample of the aforementioned location-based games and social software applications with detailed characteristics. The list of applications is not extensive, as not all the applications we mentioned in this review are included, but presents a selection of seventeen, which we identified at the time of writing as most interesting and representative of particular trends. These examples were selected from both industry and academia, primarily games or social applications, as well as on the basis of specific areas of interest and innovative features, such as elements of collaboration, user participation in public authoring and other.

The categories are displayed in two parts of a table for these same location-based games and social software. The first part of the table (2.2a) includes aspects that are specifically social, encouraging presence awareness and communication with other people, whether virtually through messaging facilities or in real world encounters facilitated by the game or match-making application. The second part (2.2b) has game specific aspects (e.g. goal) and features that are related to location itself (e.g. the facility of providing some kind of user input on a location, like the recommendations in MobiTip). Let's look at those aspects in detail.

The very first attribute is presence and as we have already discussed presence can be communicated in a minimalist way, along the lines of 'Jo is nearby' to more explicit and visual displays, such as being able to see your collaborators on a map in Catch Bob and their direction of movement. Most of these examples as we see give some kind of presence information about other users even if it's just a simple notification about their proximity. Textual interface and visual interface refer to the way information is presented to the user and how the user interacts with the device. Some applications on a mobile phone are purely text based, like in Botfighters where all the game communication is based on SMS. Other applications provide a more visual interface, like Mogi, where the user can see icons and states for other users in a mobile phone interface developed in Java. As we see most games do not have profile match-making, because this is primarily a feature of dating, friend-finder and other social networking applications, like ImaHima and Serendipity. The same accounts for 'degrees of separation', which suggests that the system links users through their network of friends and that there are various levels of acquaintance allowing a user to expand their social network, immediate friends or friends of friends and so on.

Most applications, social software and games, provide some kind of communication facility, most commonly messaging, with other users/players. Social software applications aim to encourage real life interaction among users by definition, but several games also encourage encounters in the real world as they involve collaboration or detect close proximity of other users, facilitating their identification. The aspect of pervasiveness relates to how much information is fed back to the user through the system automatically as opposed to the user requesting it. So we have strongly pervasive applications, like the Majestic game, where the whole narrative evolves through messages and communication fed to the player through a combination of media, without him or her being able to control the 'where' and 'when' this communication is received. Not so intensely, but yet pervasive, are the social software applications like Social Net that send messages with suggestions to users to introduce different people based on their profiles. The 'Uncle Roy' game is pervasive from a slightly different perspective: it requires players to give personal and contact details away as part of the game, so it challenges their privacy by blurring the fictional reality of the game with their real personal information.

Table 2.2a Categories of social characteristics

Applications	Presence	Textual interface	Visual interface	Match-making profiles	Degrees of separation	Users communicate with others	Encourage real life encounters	Pervasiveness
FLIRT project	✓	✓		✓		✓	✓	✓
Botfighters	✓	✓				✓		✓
Samurai			✓					
Mogi	✓		✓			✓	✓	
Conqwest		✓				✓	✓	
Majestic		✓						✓
Uncle Roy	✓		✓			✓	✓	✓
Catch Bob	✓		✓			✓	✓	
Seamful	✓		✓				✓	
Dodgeball	✓	✓		✓	✓	✓	✓	✓
Socialight	✓	✓		✓	✓			
ImaHima	✓	✓	✓	✓	✓	✓	✓	✓
MobiTip	✓	✓	✓		✓		✓	✓
Social Net	✓	✓		✓	✓	✓	✓	✓
Serendipity	✓	✓		✓	✓	✓	✓	✓
Urban Tapestries			✓					
Plazes	✓	✓		✓		✓	✓	

Table 2.2b Categories of game-related and location-related characteristics

Applications	Game goal	Collaboration	Users input on location	Ownership of place	Connect to web	Users create trails	History of interactions	Explicit location (maps)	Implicit location (proximity)
FLIRT							✓		✓
Botfighters	✓				✓				✓
Samurai	✓				✓				
Mogi	✓	✓			✓				✓
Conqwest	✓	✓		✓					
Majestic	✓	✓			✓				
Uncle Roy	✓	✓			✓			✓	
Catch Bob	✓	✓				✓	✓	✓	
Seamful	✓	✓						✓	
Dodgeball					✓				
Socialight			✓				✓		
ImaHima					✓				✓
MobiTip			✓		✓		✓		✓
Social Net							✓		✓
Serendipity							✓		✓
Urban Tapestries			✓			✓	✓	✓	
Plazes		✓	✓	✓	✓				✓

All the games here have a game goal, while this is not applicable to the other applications. Most of them also involve some kind of collaboration either between players in the street and online players (Uncle Roy) or between different teams claiming city space (Conqwest) or to achieve a challenging goal (Catch Bob). Socialight and MobiTip are the only mobile social software applications in which users provide some input on a location, such as a restaurant review. Urban Tapestries and Plazes are primarily focused on public authoring, so user input on different location is the main feature in these. The Conqwest game and the Plazes public authoring tool are the only examples with a notion of 'ownership of a location': users claim city territories in Conqwest and declare their own 'Plazes', networked locations they discover. About half of these examples have a web component as part of the game or system. Community websites and bulletin boards are not considered web components here, only those websites that the user has to subscribe to in order to play the game or receive information about people that match their profile. So, for example, in order to play Botfighters, a player has to build their robot online and choose weaponry etc. In the Mogi game, online players give advice to players in the street on where to collect items, so there is collaboration between people in the city and people online. There are two applications in which users create trails that are visible to other users, indicating their movement in space: Catch Bob, where players indicate their locations and trajectories on screen and Urban Tapestries, which creates trails that link each user's posts from various locations in a thread. The history of interactions refers to interactions between users that are being recorded by the system, including implicit and possibly unintentional ones, for example the number of times two devices have been detected in proximity. It also refers to the history of interactions of the user with the system, like how many tips they have posted to MobiTip and how these tips have been rated by others. So the history of interactions is not significant for all of the displayed applications, but it is certainly more important for location-based social software. The last two aspects highlight two different ways of communicating a user's location: explicitly, with geographic maps and people being represented as dots or avatars on a map or implicitly, by giving some (more or less fine grained) proximity information of other people in relation to the user. The text-based applications rely on proximity communicated verbally (e.g. 'Sally is 400 meters from you') and as we see only a few, mainly academic research projects use real map displays on handheld devices.

This categorization gives an overview of features based on examples of location-based games, social software and public authoring tools. We see that there is a strong social aspect in a lot of these; not only social software, but games as well encourage communication and real life encounters among players. This is one of the most fascinating aspects of such ubiquitous applications. We see that the notion of presence, the variants of which we will discuss again in the next chapter, is evident in most of the examples here. Presence, like proximity can be communicated with just a text message and a lot of commercial applications rely on this kind of SMS-based interface. As new applications are being developed, this categorization can be expanded and developed further. We believe it is useful to consider all these various applications as a whole, independently of whether they are classified as games or social software or something else. They are part of a ubiquitous space, creating opportunities for people to communicate, collaborate and explore the urban environment in novel ways.

Rheingold asks (2004): *How can we begin to think about a future in which cities are swarmed by constantly shifting populations of ubiquitous communicators whose devices weave ad-hoc mesh, networks of mobile communication devices?*

In addition we ask: how can we design for these ubiquitous communicators? One of the fundamental considerations for mobility is the focus on local information needs, which in a way contradicts the conventions of Internet information retrieval, where searches start from a general, global view, to more specific content (Sacher & Loudon, 2002). This does not imply that global information is irrelevant to mobile users, in fact, the aforementioned examples combine both. It simply suggests that mobile users are likely to be interested in local content first.

Another challenge is blurring the boundaries between the physical world and virtual information. One problem is how to bridge the richness of the real world with the limited display and interactivity of a mobile device. But this is a creative problem: designing for interaction with the environment and other people while being on the move has opened up opportunities for the designers of the games and social applications mentioned in this section. New elements come into play. For example, using streets as a game board not only questions the definition of gaming, but also brings new nuances and level to the production of meaning in urban space (Sotamaa, 2002). If the mobile gaming ideal is to free players from the chains of time and place, location based gaming on the contrary operates through creating new meanings for familiar locations. Then again emotions, memories and perceptions attached to places can affect the game play, for instance in some cities there are certain no-go areas.

Designing for ubiquitous communicators also means designing for short, often interrupted interactions. It is within the nature of the medium not to allow long interactions. Commuting time can provide an appropriate time for play, social interaction and exploration of the surrounding environment, but these experiences can be interrupted at any time. Unlike online massively multiplayer games, it is doubtful that mobile players would perform complicated actions through their mobile phone. This in turn changes the whole relationship to playing itself towards a more casual mode of play, which attracts a much broader audience than that of hardcore gamers and can encourage truly massive participation.

2.2.5 Research framework

Based on literature reviewed here and our discussion on Flash Mobs, the parameters in the concept map in figure 2.7 represent the main research issues this thesis aims to explore. The parameters are investigated in the context of both purely online and mixed reality interactions, with the exception of the parameters of *communication* and *behaviour in public*. The thesis explores presence-based *communication* in online environments, aiming to find out how visual communication can facilitate the emergence of spontaneous behaviours and collaboration. At the other end of the map, in the context of mixed reality interactions, we aim to explore people's *behaviour in public* spaces and particular issues that were highlighted in the earlier discussion on Flash Mobs: the willingness to perform certain acts in public, the sense of embarrassment or the absence of it, conformance to social conventions and people's awareness of the surroundings while participating in a mixed reality game. In relation to this, the thesis addresses notions of *social acceptability* in both online and mixed reality contexts: are the unexpected, unpredictable behaviours we are interested in accepted by the people participating in the experience and the broader social environment?

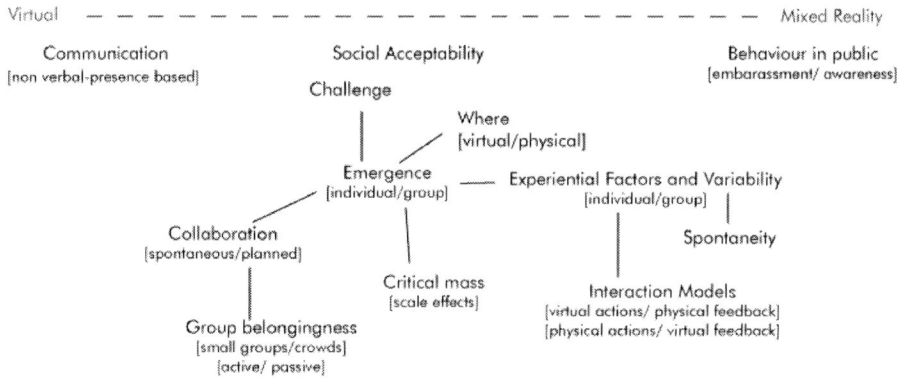

Figure 2.7 A map of the research framework, illustrating the parameters under investigation and their interrelations.

Emergence is at the centre of the concept map and relates to other parameters, such as the *where*, whether emergent individual or group behaviours are expressed in the virtual (online) or in the actual physical world, motivated by the virtual interaction (mixed reality). *Challenge* can result in emergent behaviour, for example Flash Mobs participants responded to a particular challenge to perform an embarrassing act in public in front of unsuspicious passers-by. Earlier discussion on play in this chapter highlighted the importance of having a challenge for motivation and the enjoyment of a playful activity. In the context of mixed reality and location-based games, the parameter of challenge includes trade-offs and decision making processes within the game. Our conceptual investigation through design in chapter 4 discusses some of these issues in attempt to satisfy the need for challenge while maintaining a sense of trust and privacy in mixed reality game situations.

Emergence influences another important parameter, the experience of social applications and games. In particular, we set out to investigate the *experiential factors* by identifying ways in which we can design for experiences that vary and evaluating the designs. *Spontaneity,* as in playground play, is related to experiential variability and the unpredictable interactions we are aiming for. The research also considers *critical mass* as a variable for emergence and we expect that changing the number of participating individuals can affect the emergent interactions as well as the experience of the game/ application.

Collaboration is another important parameter under investigation. Following from the discussion in the literature on emergent collective actions and large scale collaborative games, we want to see how much 'structure' needs to be in place in the design of a game to facilitate emergent collaboration, both in online and mixed reality contexts. Do we have to design specific features in an application for people to collaborate or can collaborative activity can be to some extent unpredictable? What forms of collaboration can we observe through play?

One of the premises of this work is the idea that a sense of people's presence can enhance the feeling of being part of a group, *group belongingness*. Central to the concept of presence based group play is the social cohesiveness factor that can be achieved when people participate in engaging social experiences online and in mixed reality situations. We aim to find out, whether in the first place, group belongingness can be achieved through spontaneous play based on presence, following with further insights into its manifestations. Group belongingness can range from simple expressions of feeling part of a group of people (e.g. along the lines of 'I check my IM list to see who else is online') to more active manifestations in playful contexts (e.g. collaboration, team loyalty). Our long-term goal is to explore this range across online/ mixed reality situations, while incrementing the number of participating individuals. Can group belongingness be achieved among large groups or crowds of people?

In mixed reality games and social applications it is very important to establish a balance between the mediated (virtual) interactions and users' attention on the real (physical) world. This is one of the primary explorations of the thesis. Because activity and awareness need to be distributed across virtual/physical reality, we need to determine *interaction models* that facilitate this kind of cross-over and enable users to engage with an experience 'in-between' mediated, virtual actions and feedback in the physical environment and vice versa.

The next chapter introduces our approach of designing for emergence, as an experimentation framework to uncover unexpected interactions and social uses of technology that can inform the design of novel interactive applications. In order to 'set the scene', we present our guiding design principles.

3. Design for emergence

3.1 A model for design for emergence

The motivating theme for this thesis is *design for emergence*. How can we design for the unexpected? Can we encourage unpredictable, social uses of technology that could inspire innovative designs?

Why design for emergence? Johnson (2001) has described how artificial intelligence scientists and game designers like Will Right who designed SimCity have understood and recreated emergent systems based on simple rules that have a complexity and life of their own. In fact, emergence is a particularly intriguing buzzword for game designers (Garneau, 2002) as they try to incorporate emergent properties in their games to enhance the user experience. Emergent interaction based on simple, high level rules that can vary result in a different experience every time, a game that is interesting to play more than once. Game design is in fact, a second-order design problem (Salen and Zimmerman, 2004) as the designer designs the rules of a game directly, but designs the player's experience only indirectly. The experience of play depends on the emergent interaction and it is not always possible to anticipate how the rules will play out or how the player will behave within the game. Because of this uncertainty, it is indeed difficult to design a game based on a simple set of rules that will generate complex, yet meaningful and engaging interactions.

Emergence is not only valued in game design, but other design disciplines too, such as computer-aided design. A property of a design that is not represented explicitly at the time of creation is said to be an emergent property if it can be made explicit (Saunders and Gero, 2001, Gero, 1994; Mitchell, 1993). Protocol studies of designers while sketching have shown that unexpected discoveries of emergent shapes can have a significant impact on the course of further design activity (Schön and Wiggins, 1992; Suwa, Gero and Purcell, 1999).

While the above studies focus on emergence during the design process itself, we are specifically interested in emergence occurring from the unintended uses of technology. We have many examples of unintended uses of technology, such as, SMS-based on-the-move coordination, hijacking Bluetooth phones to communicate with random strangers in short range, time-shifted radio broadcasts from iPods (podcasts). GPS-art, drawing pictures by tracing your own walking paths is another example of using technology in an unpredictable way, such that the system designers did not intend when they developed and launched GPS in 1989 (Pryor and Wood, 2001). Often these unintended uses of technology repurpose the design towards a new direction, incorporating ludic values. Unintended use as a design 'approach', can also be seen as a way to open up for the public to take part in the shaping of a public sphere, the cyberspace (Stolterman, 2002). While there is an economic interest in designing for the unexpected (e.g. SMS revenue, killer apps), it is also beneficial for the design process as well, as users can become part of the process by pointing out new design directions and providing inspiration for designers.

Andersson, Broberg, Bränberg, Janlert, Jonsson, Holmlund, and Pettersson (2002) suggest their framework of emergent interaction systems as a means to design for emergence. They define such a system as an environment with a number of actors who share some experience/phenomenon, and whose behaviour is significantly influenced by a shared feedback loop picking up data from the individuals and their actions. The immediate effect of emergent interaction can enhance the individual experience. Their goal is similar to the one our own research aspires: to increase or enhance social life with participation and engagement and to provide new ways of forming acquaintances (Andersson et al, 2002). They have formulated an emergent interaction system model (figure 3.1) that can be applied to a wide range of areas, from exercising events, public space, arena events to events taking place in the home environment, in a professional/educational or collaborative art context. However, since the model has not been tested with any prototype, we cannot be sure about whether it really works for any of these contexts, whether it is possible to design for emergent interaction with this feedback-based system approach.

Emergence is a very popular term nowdays, but it can have more than one definition. For example swarms, flocks and large scale interaction like Carpenter's massive auditorium Pong game, discussed in chapter 2 are all macrobehaviours emerging from a combination of individual interactions. Here, we consider emergence in the broadest sense possible, both as a collective but also as an individual phenomenon. If a user of a particular novel technology discovers a new, unexpected way of using it, then possibly other users of the same technology would adopt it too.

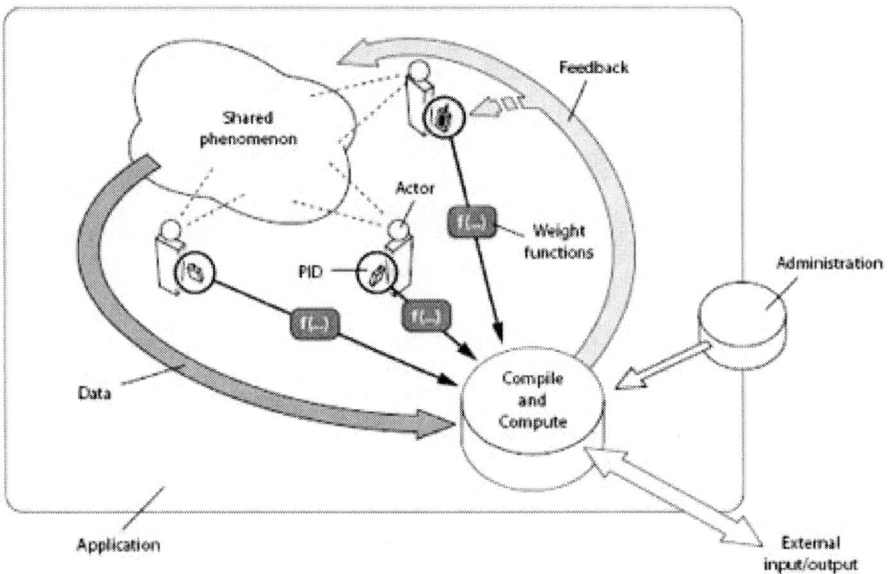

Figure 3.1 Andersson et al (2002): a functional model of an emergent interaction system. The basic components are a) the shared phenomenon, like a synchronous experience or shared reality, as being in traffic, a sports arena, a theatre etc. b) the actors, participants in the shared phenomenon and c) the application, a system set-up dedicated to some purpose or event, e.g. theatre, interactive art etc.

Considering the interaction design process for the development of new, innovative products, Preece, Rogers and Sharp (2002) suggest a model drawing from their observations of interaction design practices and from previous software engineering and HCI lifecycle models. With this they describe an iterative design process (figure

3.2): based on feedback from the evaluations, the creative team may return to identifying needs or refining requirements, or it may go straight into redesigning. Implicit in this cycle is that the final product will emerge in an evolutionary fashion from a rough initial idea through to the finished product. The only factor limiting the number of times through the cycle is the resources available.

In our research work we focus on the flow between design or redesign and evaluating an interactive application. The long-term aim is to understand how unexpected uses from the deployment of a new technology can drive its design and whether it is possible to incorporate those in the iterative design process. For this reason we try to identify both the design elements as well as the external factors that facilitate emergent social behaviours and unpredictable uses of technology.

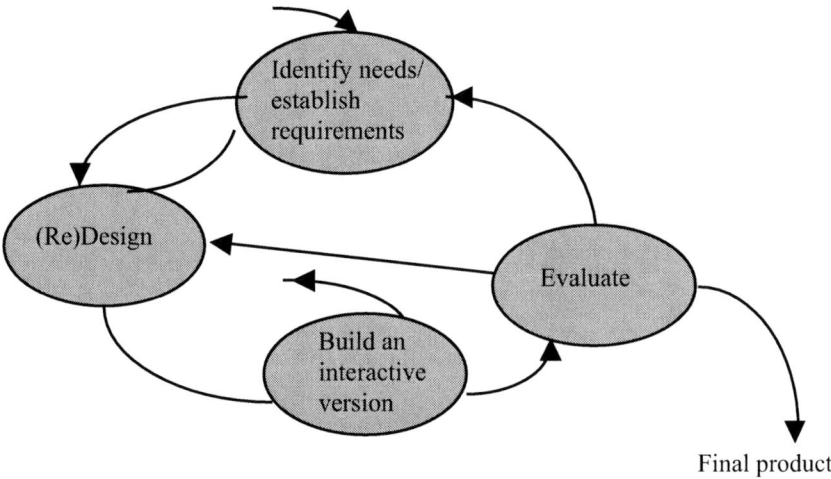

Figure 3.2 The iterative design process by Preece, Rogers and Sharp (2002)

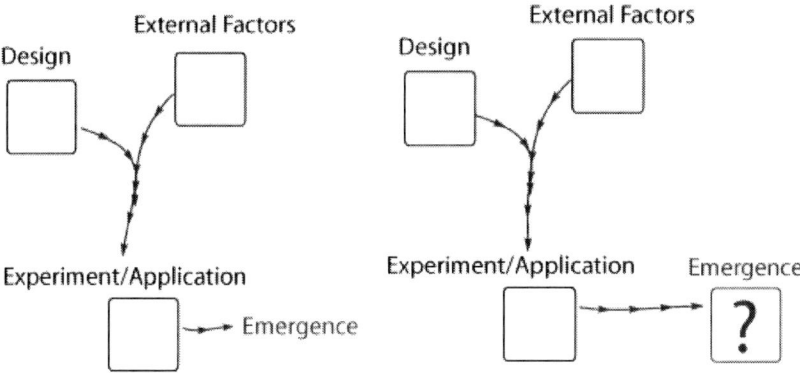

Figure 3.3a Emergence, as unpredictable behaviours and uses of technology, comes from a combination of design and external factors through the deployment or experiment with an interactive product. Our aim is to study emergence in a way that makes it as important as the interactive application itself.

The model in figure 3.3 (a and b) describes the focus of this thesis, aiming to capture the core aspects of emergent phenomena from unintended uses of technology. The model brings emergence to the forefront. Rather than a side effect that occurs from the deployment of an innovative technology, emergence becomes of a primary interest in our research work.

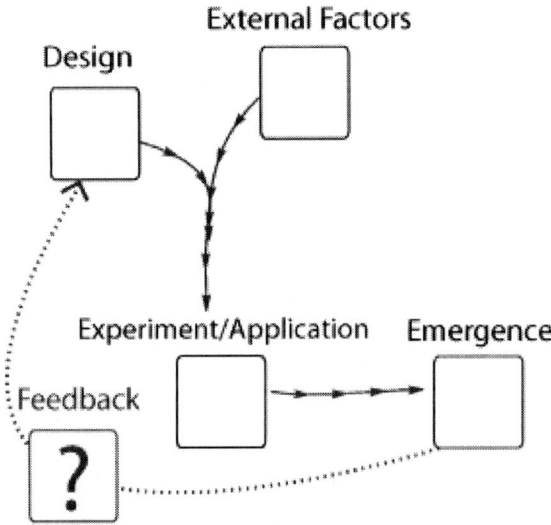

Figure 3.3b The long-term goal is to try and integrate lessons learned from emergent user behaviours in future designs.

Consider the emergence of SMS messaging in light of this model. The design is very simple and minimal. The Short Message Service is limited by a maximum of 146 characters of text and input that is quite hard for first time users as each button press on the mobile phone corresponds to several letters, unlike Qwerty keyboards or voice, which was what we used mobile phones for. Unlike voice services or online chat, SMS is more asynchronous and less immediate. So the design is quite limited, however the external factors that led to the huge success of this service are quite significant and they are primarily social and psychological. Playfulness and flirting are communicated with short clever messages among young people. Moreover, SMS is less intrusive than a phone call so sending a message to someone in certain situations is more polite and appropriate, so it took off in the business environment too. Also, the sender feels more comfortable by not having to talk to the receiver and interrupt any other activity, but can still give a signal of awareness, much like confirming one's presence: 'I am here'. For example, the exchange of simple 'goodnight' text messages creates a sense of connectedness (Rettie, 2003), even though no further communication takes place. The extensive use of mobile phones among teenagers and the fact that sending a text message was cheaper than making a phone call were also external factors that influenced how SMS was being used. So from the deployment of a limited service, only as an addition to the main voice service on mobile phones, what emerged was an entire new language, along the lines of 'c u l8tr' and new communication practices for keeping in touch with other people. Eventually the success of SMS opened new directions for data services on mobile phones and more enhanced messaging services were designed: EMS and MMS. The integration of data services in a device that was primarily used for voice communication, like the traditional phone, is emergence in its own right and today's mobile devices are examples of technological convergence, with both telephone and computer features. The diagram in figure 3.4 illustrates this design process based on the emergent properties of SMS usage.

Now consider how this model applies in the context of this thesis. One of the key questions this thesis is trying to address is how (if at all) we can design for the emergence of spontaneous social behaviours and play. One part of the thesis concentrates on emergent behaviours and social interaction online with a case study of a multiplayer game and another part explores how spontaneous individual and collective behaviours emerge in the real world, through the use of mobile technologies. In both case studies, the research focuses on how play, based on very simple game rules, can lead to more complex interaction, emergent cooperation and collective behaviour.

The principles in section 3.2 describe our approach for design for emergence, following our observations of various interactive applications discussed in chapter 2.

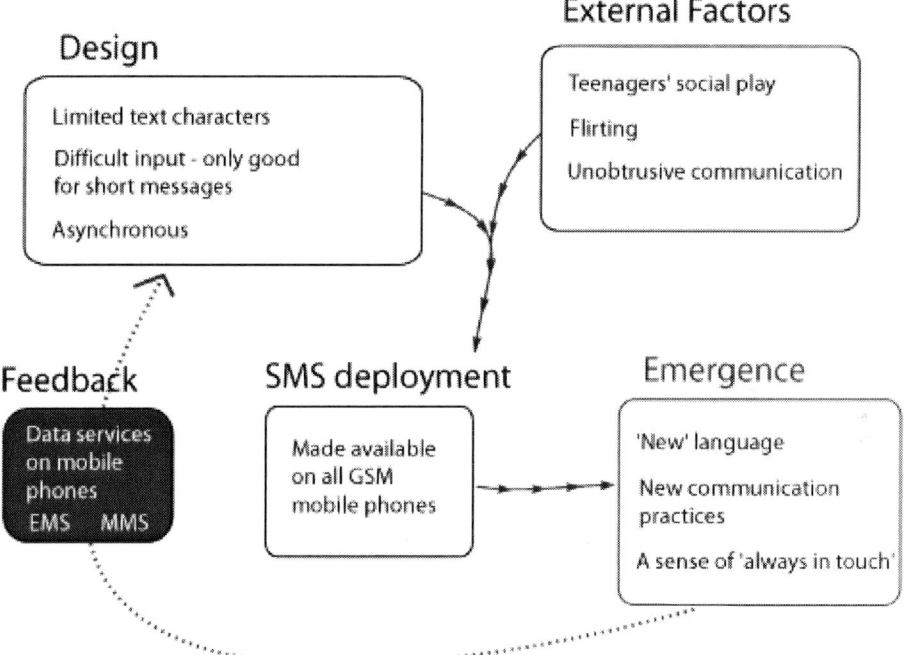

Figure 3.4 The model of emergence for SMS messaging

3.2   Design Principles

3.2.1   Presence is symbolic

   a) Presence can be rendered symbolically in an online context

In the literature review of this thesis in section 2.1.1, we discussed our definition of presence as simply knowing, *being aware* of other people's existence. Our diagram of typical presence 'states' found in Instant Messaging applications illustrated various aspects of a person's presence, such as availability, activity, location etc. These states show how presence is conveyed symbolically in the online world.

We believe that the sense of others' presence online is both *necessary* to achieve communal impact and *sufficient* to induce the appropriate sense of 'feel good' or 'buzz' in others. The social software presence visualizations discussed in section 2.1.1 are valuable contributions towards this direction. In the context of spontaneous play online we are aiming for, similar, game-specific 'states' can facilitate communication among players by providing visual cues, without the need for explicit oral or text-based communication. Our belief is that simple visual cues and states, such as colour coding to indicate team membership and direction of movement can facilitate coordination and collective playful activities online, particularly when many people participate at the same time and sending a text to everyone saying: 'let's go to the left' for example is an inefficient way to communicate.

b) Presence can be overlaid symbolically in a mixed reality context

We know that in the online world, an individual's presence can be conveyed symbolically via the display of meaningful *state* information (e.g. availability, activity, location, team identity etc). In ubiquitous computing environments, presence becomes a richer concept creating a hybrid space of mixed reality, as virtual (symbolic) presence can be combined with physical (real) presence, through information about a person's location or proximity to us. Location information introduces a space of opportunities for mixed reality interactions, also exemplified by the emergence of a whole new genre of location-based games, dating services and social software discussed in paragraph 2.2.

Table 3.1 Minimal, abstract and explicit presence, based on location or proximity information.

Presence	Location information	Display	Design
Minimal	Much like online/offline, 'you are there or you are not there'. The information is that someone is in vicinity	A name (or nickname) of a person, possibly with a vague indication of how close they are (e.g. same cell or postcode)	Textual: SMS messaging, chat
Abstract	You receive information about a person being near and their proximity to you	A radar view for example, where you can see also the direction to which the person is located and an estimated distance from yourself.	Classic radar views as used in multiplayer games
Explicit	Exact location information.	Typical geographic map display with the location of the person displayed on a specific point.	GPS device interface, maps

In the online world there are variable levels of presence depending on how much state information is displayed, ranging from a very basic online/offline state, to richer context, including information about a person's availability, activity, mood, device capability etc. Similarly, in a mixed reality situation we can also have variable levels of presence, depending on how location information is displayed and how much is revealed. Drawing from text messaging, game design and traditional location displays via maps on GPS-enabled devices, table 3.1 illustrates three possible ways to communicate a person's presence based on more or less explicit location information.

In those three cases, presence information ranges from minimal to explicit. Depending on the type of game, or location-based application we are designing, a higher or lower level of granularity can be appropriate. More variations are possible, so for example a portable device interface could show someone's name as well as their distance from the user, but no direction. This would be another form of abstract presence, yet even less explicit than the directional radar view used in games. In our work, we aim to find how much presence information is appropriate in a mixed reality concept, to stimulate imagination and play, while maintaining an awareness of the surrounding environment. Somewhere between minimal and explicit presence a balance should be achieved for most location based games and social software, depending on the required granularity.

Our goal is to design a game based on symbolic presence states superimposed over physical presence, in a way that both *enhances* the feeling of being part of a parallel virtual experience as well as *facilitates* the emergence of spontaneous social behaviour in the real world.

3.2.2   *Large scale is important for emergent interaction*

Rather than thinking of very large numbers of participants as something which might 'flood the network' or 'slow down the software', we prefer to look at large scale as a key enabler and therefore an important starting point in this research. At the very least, we need to bear in mind that certain interactions, typified by crowd phenomena such as swarming, Mexican Waves and Flash Mobs (London Flash Mob Website, 2003), can only happen at large scale. The Cloudmakers example described in section 2.1.2 (7,000 online puzzle solvers who attempted to solve real world challenges) illustrates the importance of large-scale for collective problem solving. Large-scale community interaction is a great challenge for collaborative systems designers and collective play can act as paradigm to inform real life group problem solving.

Earlier in chapter 2, we also mentioned the Mexican Wave as an inspirational example of spontaneous, emergent group behaviour in the real world. Such spontaneous synchronous interaction online has yet to be explored and we are interested to find out what the equivalent of the Mexican Wave phenomenon would be in the online world. We expect that, much like the real world Mexican Wave, an emergent collective behaviour online would require a critical mass of participants to be present and for this reason we need to try and involve as many people as possible. A significant part of our work has been focused on how we can design for the emergence of 'crowd' behaviour online. Although the empirical studies of this thesis necessarily investigate this only with small numbers, in the longer term we are interested in applying the lessons we learn here to the design of much larger environments, facilitating emergent play among hundreds or even thousands of people. Therefore, the applications we used to obtain our results have been designed with the future potential of large-scale in mind.

It is well known that nowadays most people carry a mobile phone, and sometimes they do not have a personal computer or any experience of using computers at all. At the same time technological convergence has led to the creation of all-in-one devices (both voice and data services). Mobile phones gradually incorporate computer specific functionality, for example Wireless Instant Messaging (Vogiazou, 2002). Therefore, possibilities for large scale engaging experiences, mediated by mobile technologies, are emerging. Starting with location based multiplayer games, we are interested to find out what kind of games we can design that would take advantage of situation-specific contexts and mobility to involve hundreds or thousands of users, or even an entire city. We are particularly interested in emergent social behaviour, enabled by mobile technologies and our experiments in chapter 7 have made a step towards this direction. As Rheingold says (2002): *Location-sensing wireless organisers, wireless networks, and community supercomputing collectives all have one thing in common: They enable people to act together in new ways and in situations where collective action was not possible before.*

3.2.3   Keep the design lightweight

This design principle comes as a natural outcome of the two principles above, the communication of symbolic presence and the interest of designing for scalability. In order to communicate presence states immediately and to promote scalability, we follow the rule: '*keep it as simple as possible*'. We strongly believe in the concept of 'complexity growing out of simplicity' as being crucial for the success of mediated social experiences, where group behaviours can emerge spontaneously through a simply designed application. The basic challenge is to provide 'just enough' design to create an environment that is conducive to forming rules and relationships rather than enforcing them. Lightweight design online, inspired by the symbolic communication of a presence state in an Instant Messaging application, such as being online/offline, can enhance the visualization of an online 'crowd'. Our studies in chapter 5 present experiments with a lightweight, presence based multiplayer online game.

The same approach has been applied to the mixed reality game experiments in chapter 7. In the context of ubiquitous computing and mixed reality experiences where attention is distributed between the surrounding environment and people and a device interface, the principle of lightweight design becomes even more crucial. Mobile technologies enable short, spontaneous and intermittent interactions while on the move (Raby, 2000) and it is precisely these playful interactions we aim for. We know that mixed reality, location-based experiences are characterized by a transition between immersive and non-immersive states (Reid, Geelhoed, Hull, Cater and Clayton, 2005) caused by either planned events or occurring interruptions such as a happening in the physical world or a system fault. The design challenge here is, unlike fully immersive computer based games, to smoothen these transitions between immersion and non immersion. Therefore, any ubiquitous game design should allow for awareness of the surrounding environment and stimuli external to the game and encourage ad hoc play. Our goal is to enhance our interaction with the social and physical world around us by adding an additional experiential layer, our superimposed virtual, symbolic presence. Because the main focus should be drawn on the surrounding environment and people rather than on the game/device itself, our design needs to enable and encourage this kind of interaction. Therefore, the lightweight design approach is necessary to facilitate awareness of both physical and virtual situations at the same time. Mobility, distributed attention and interrupted interactions while being on the move, all point towards this direction.

In this way the design acts as a prompt for the real world, real people being physically close. Drawing on the design of presence enabled applications like Instant Messaging, where communication takes place with simple, symbolic presence state changes, ubiquitous, mobile interactions can also be based on symbolic states in the form of alerts which point to the real world (e.g someone is really close to me now). This is the reason we favor rather abstract, implicit representations of presence information, along the lines of communicating a sense of proximity rather than accurate location on a map interface.

Our empirical studies in chapter 6, illustrate how spontaneous behaviours in the real world can emerge through a game designed as a prompt for the real world, real people being physically close.

3.2.4 By employing affordances users understand and can extend the design

Reflecting on guidelines for design for emergence, a first question that comes up is how can we finalize as little as possible, to encourage people to extend a design through their explorations and yet provide enough context for an engaging and meaningful experience. Norman (2002) has coined the term *affordance* to refer to the perceived and actual properties of an object that determine how the object could possibly be used (e.g. knobs afford turning). Affordances provide strong clues to the operations of things.

Similarly, a rectangular box in a chatroom affords typing some text into it, because of the context of use and our familiarity with conventions of computer interfaces. The design of non-physical, digital objects can also suggest associations with objects and experiences of the real world; this can influence the way people use and relate to the digital objects. By employing affordances in our designs, such as using simple metaphors, related to people's past experiences that can be re-interpreted or subverted, we aim to provide for creative play. By allowing people to explore strategies and cooperation practices themselves, without incorporating them in the actual design, but encouraging team play at the same time, we expect to see spontaneous individual and collective behaviours emerge.

The grounding in familiar visual metaphors also enables people to focus on the content of a situation, avoiding incurring cognitive overheads to make sense of what is shown. In 'Mapping Cyberspace' (2000), Dodge summarizes various map types with different meanings and metrics. Among other things, the author mentions *real location maps*, *office maps*, *logic* or *schematic maps*, *mood maps*, *interest* or *topic-centered maps*, and *project* or *progress maps*. We know that the presence of large numbers of people can be represented visually using density plots on maps (Dodge, 2000), and in very compelling ways such as the NASA Earthlight maps (NASA, 2000) which reveal the most densely populated areas on our planet via a stitched-together global panorama of night-time satellite photos showing city lights . Social network visualisations are good examples of depicting the rich variation in relationships among people (Freeman, 2000). Another interesting visualisation is 'The World as a Blog', a website which combines real time blogging activity with the blogger's actual location represented as a dot on the planet (Maron, 2003). The HitMaps visualisation of Komzak and Eisenstadt (2005) provides a highly scalable view of the locations of tens of thousands of visitors to blog sites in a tiny 'gutter' display. Such scalable visualisations provide a familiar grounding for experiencing the presence of large numbers of people online.

The principle of providing the affordances for people to be able to relate to an online environment and to experiment within it, suggests that we need an appropriate visualisation of the overall virtual space and the events happening within it. In this way, peripheral awareness of other participants will be supported, so that if a collective and possibly serendipitous activity emerges online, others will become aware of it and might join in or respond to it.

Considering mixed reality social experiences, we need to provide affordances to support the transitional interaction between a virtual, superimposed situation and the physical world. In this case, providing an overview of a whole virtual environment is less useful, as there is only a limited amount of information a user can comprehend when being on the move, using a device with a small display, having distributed attention etc. When being mobile, local information (e.g. who is within 100m from me) is of much more relevance than global (e.g. who else is in the same city or country). One approach we pursue here is to use appropriate metaphors that relate to peoples' past experience and link the virtual with the physical world, in a way that the metaphor strengthens the relationship between the two 'worlds'. By blurring the boundaries between the virtual and the physical we aim to encourage a new genre of emergent behaviour and interactions, motivated by a virtual context, but expressed in the real world.

The next chapter illustrates our design attempts for emergent collaborative social play, both online and mediated through mobile technologies, in a series of storyboards.

4. Early design sketches for design for emergence

In this chapter we describe our initial concepts for multiplayer games aiming to facilitate some kind of emergent behaviour. This chapter illustrates the conceptual phase of the research, which included several brainstorming sessions, followed by idea generation through drawing and interface design. Although there is an extensive literature on the role of brainstorming in group processes and productivity (Isaksen, 1998), the role of brainstorming as a design research tool has not been sufficiently documented. It is however a widely accepted creative problem-solving method for design. Green and Bonollo (2004) report that new systematic design methods were first introduced in the 1960s and were applied in certain fields of design practice, including engineering, industrial, architectural and urban design. During the same period, the techniques of creative engineering and brainstorming became more widespread and these provided some bases for idea generation. Brainstorming is used in the conceptualization phase (Green and Bonollo, 2004) of the design process. This phase has been important for the thesis and the research issues outlined in the previous chapters have been treated as parts of a broader design challenge. This challenge is to create engaging, participatory social experiences for large numbers of people, mediated through technology. The ideas included in this chapter have been generated through brainstorming (both individual and in a group of 2-4 people), following the principles outlined below.

A central principle involved in brainstorming has been described as 'deferment of judgement', which means that judgement is postponed during the process. Osborne (1953) has defined the following four fundamental principles for brainstorming: a) *Criticism is ruled out*. Adverse judgement of ideas must be postponed until later. b) *Freewheeling is welcomed*. The wilder the idea, the better; it is easier to tame down than to think up. It is desirable to share really wild and unusual ideas. c) *Quantity is required*. The greater the number of the ideas, the greater the likelihood of useful ones. d) *Combination and improvement are sought*. In addition to contributing ideas of their own, participants should suggest how the ideas of others can be turned into better ideas; or how two or more ideas can be joined into a new idea.

After the brainstorming session, the generated ideas were sorted and evaluated. The most favourable ideas were then further developed through the drawings and interfaces included here.

We use our model for design for emergence to show how these concepts address emergence and identify fundamental design characteristics and external factors that can be influential with regards to the emergent interaction. These are exercises to speculate what kind of emergence we can expect and how that would fit in the design process. Concepts have been developed for a range of different interactive technologies: starting from the online medium (section 4.1), where interaction is computer based, then considering the use of mobile technologies and their limitations (section 4.2), to finally address the challenges of ubiquitous computing and mixed reality interaction (section 4.3), where the game design takes advantage of the physical environment as the virtual experience interweaves with the real world. Although we did not develop and did not experiment with any of the games mentioned in this chapter, the ideas and themes behind them are very relevant to this thesis and therefore included here.

4.1 Online games

One of the first challenges this research aimed to address was: can we design a game for potentially large numbers of people, based on their mere presence? The notion of presence states, as in IM applications, has been influential henceforth, and this becomes evident in the concepts. When using Instant Messaging, a person's availability or state of activity is indicated by symbolic icons. One of the key design challenges in our experimentation is the use of symbolic presence in a new context; to see how visual information (e.g. colour, movement or shape) can communicate a person's actions and social behaviours within a game. We are interested to see to what extent changes of 'presence' states indicating behaviour and intention can be effective. Research in group collaboration tools and chat room design has proven that abstract representations are very scalable and efficient to use (Erickson et al, 2002, Viegas and Donath, 1999). Our aim is to see whether visual and symbolic communication can be effective in assisting players to coordinate their activity. The design challenge is to represent presence information in a meaningful way, within the context of participatory activities.

In order to encourage emergent behaviours, we decided to develop cooperation challenges. Social psychology theories have provided some interesting background on group behaviour and cooperation. For example, the Social Identity Model of Deindividuation (SIDE) (Lea et al, 2001) explains strong group cohesion effects in visual anonymity situations in computer-mediated communication. The first game concept addressed the shift of focus from the individual to a group and aimed to explore whether we can enhance a feeling of *group belongingness*, particularly as the numbers scale up.

This game concept aimed to promote large-scale interaction and emergent swarm behaviours in a maze environment, based on the theme of the classic arcade game of Pac-man. In this game we reversed the idea behind Alexander Mintz's panic simulation study (1951), described in chapter 2, to inspire a new form of collaborative crowd behaviour. The theme of clustering, flocking as a cooperation strategy is important here and it comes back again in this thesis as an emergent behaviour we have been aiming for with our designs.

In this maze-like hunting game players cluster around an exit in order to open it and push obstacles together out of their way. The challenge is based on group division: one group is the 'hunters' (purple scissor-shaped icons in figure 4.1), much like the ghosts in Pac-man) and the other is the 'runners' who try to beat the first group by cooperating in different ways. The 'hunters' try to capture the 'runners' whose aim is to find the right exit and manage to escape. Each player is represented by a little figure and an arrow, indicating his/her direction of movement. When a player moves alone in the maze world, his/her movement is very slow and sometimes he/she gets slowed down by other players moving to opposite directions or even by obstacles. The key point is to link to other players of the same group that move to the same direction and move altogether, like a crowd of commuters in the underground for example. By clustering and moving together, players of either group can move faster and gain strength. For example, the purple 'hunters' can chase individual 'runners' because they can now move faster and capture them, whereas if 'runners' flock, they gain 'group power', move faster to avoid 'hunters' and they can also push obstacles out of the way if there is a sufficient number of players in their flock. If a group of 'runners' (e.g. at least 10) cluster together on one spot, they can create a temporary safe zone, a protective shield that makes them immune to hunters' attacks for a certain amount of time (note the group surrounded by a square, protective border in figure 4.1).

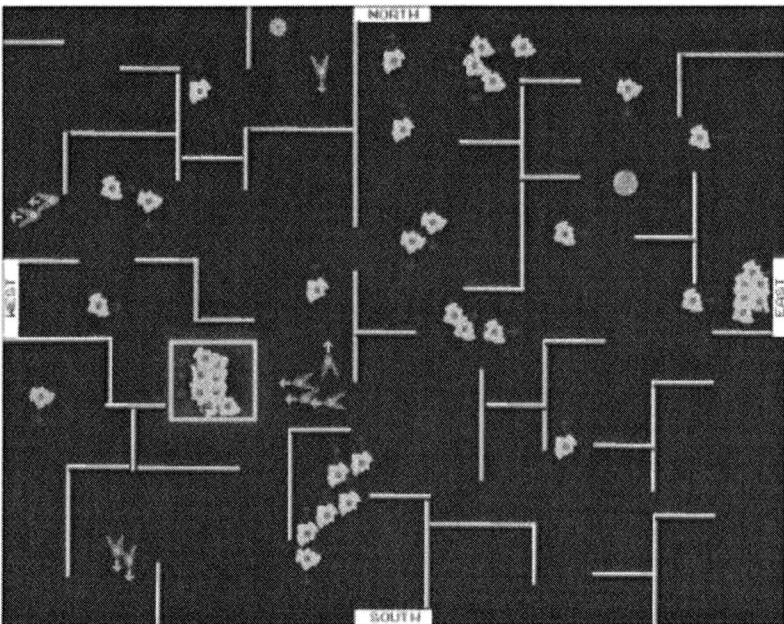

Figure 4.1 A maze-like hunting game

There are four exits out of the maze, north, south, east and west. At least one (or more) of them is a dead end, therefore the aim of the runners is to find the right exit to escape from the maze. In order to open those exits, a group of 'runners' need to push against them. If they succeed they earn points and status and they re-enter the game as hunters. To help them find the right exits, clues with the icon *i* for information appear randomly for a few seconds and can be picked up by players, if they happen to be on their way. A player who has picked up an information clue, will get an *i* presence status for a limited amount of time – this will make others notice that this player is a carrier of information about a correct exit. Players can send short messages to each other exchanging these clues. Another interesting parameter to explore here is 'rumour' spread; how people exchange information and how this *i* status feature affects their behavior in the game, as well as the behavior of the others around them.

Performance feedback and reward elements could be introduced to add variety to the game. For example, if a player picks up a strength element, he/she can push obstacles and apply more strength to the exit. Or if players succeed in moving along with a group of other people and pushing obstacles together, they could get a reward for cooperation. Player profile information could be included to add a social dimension to the game and enable people to experiment with their fictional identities.

The novelty of the Maze game is in a) the fact that the emergence of spontaneous collective behaviours depends upon having a critical mass of users and increasing the number of participants can produce variable user experiences and b) that presence 'statuses' are introduced within a game context (e.g. indicators like direction or *i* for information) to facilitate ad hoc coordination and scalability.

This idea generation process informed our research framework outlined in 2.2.5, particularly in relation to the parameters of *challenge, collaboration* and *emergence*.

- *Challenge*. The Maze game poses a number of challenges to the players, whether 'hunters' or 'runners' (e.g. capturing other players, moving obstacles, finding the right exit, creating a protective shield with others), which require collaboration. Other challenges are individual (e.g. find information clues), so the game aims to address different player styles and enrich the *experiential factors and variability*. The outcome of the game is quite unpredictable as it depends on the parameter of *critical mass*, i.e. how many people will be participating in each team as well as the players' ability to collaborate. At the same time, there seem to be too many challenges for the spontaneous, presence-based, playground type of interaction we are trying to design for.
- *Collaboration*. The game clearly encourages player collaboration to address the challenges. Most of it is planned collaboration though, assuming that the participating individuals know for example that they need to cluster in order to move an obstacle. However, some of these interactions could be part of the game but not made known to the players from the start, to allow discovery and spontaneous collaboration. It would be interesting to see if any forms of unplanned collaboration emerge, that are not part of the game context.
- *Emergence*. This is almost an exercise for flocking behaviour. While this kind of emergent behaviour is particularly intriguing for us, it seems that there is already a bit too much planning and structure to allow for truly unpredictable behaviours to emerge. Maybe other collective behaviours could occur in the game, but we would only be able to find out through implementing the prototype and carrying out user trials. This game would be useful to see how the player number can influence interactions among swarming groups and whether self-organisation online is at all possible.

Deploying the design for emergence model for this concept (figure 4.2), we see that presence indicators, challenges requiring collaboration (like finding the right exit) and the interaction of clustering to gain strength, speed or protection in the context of a scalable maze game are the main design elements. Depending on the numbers of people playing on each colour coded group the outcome can be different every time. Also, integrating IM elements in a game like this, like messaging and contact lists can bring interesting social interactions as people are likely to get to know others through playing the game. This experimental game would help us identify the kind of collaborations that are possible and estimate how many users can partake in each emergent interaction. For instance, would it be possible for 10 users to move together and 'push' an exit in the game without any leadership or top level coordination?

If the answer is yes, then these results could feed back into the design process to add more dynamics in collaborative play: we could allow for more complex interactions; groups of 'runners' teaming up to become 'hunters' (similarly to the original pac man game where pac man can chase the ghosts for a limited time). The game could have secret 'tweaks' changing the game world itself that could only be discovered if a certain number of people performed an act together. Beyond the concept of this particular game, emergence could inform the design of a different type of experience. In principle, we could design a multiplayer educational simulation of a living organism or biological processes, such as the immune system's antibodies fighting against viruses. Players would undertake different roles and learn about management of resources and coordination and biology at the same time. If, for example, we found from the maze game that groups of 5 people can coordinate more efficiently, then the educational simulation could assign a role to groups of 5.

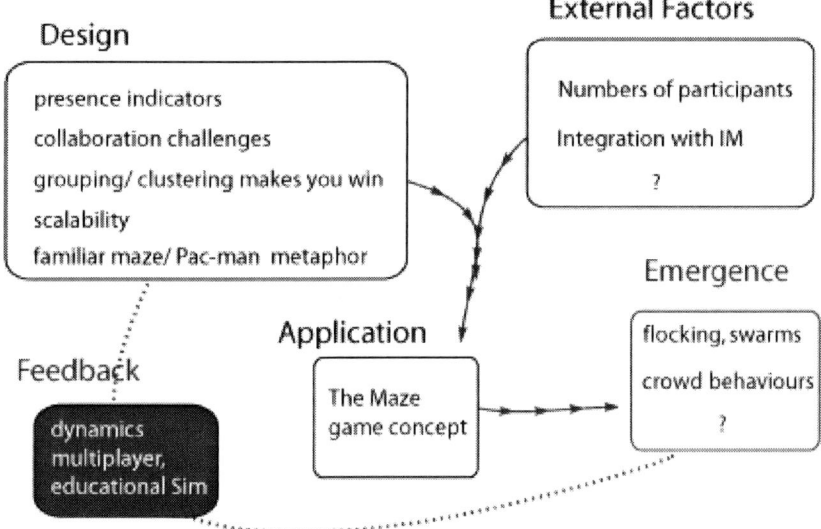

Figure 4.2 Speculative design for emergence with the Maze game

Overall, the Maze concept provides good metaphors and paradigms to explore design for emergent group and collective behaviours in an online gaming context. If deployed on a large scale, it would provide a useful framework to study self-organization online and define the required *critical mass*. By varying certain communication parameters (e.g. how much players know about what they can do from start or whether messaging is available or all *communication* is visual), we could find out how these variations influence emergent collective behaviour (flocking, swarms) online.

4.2   Pixeltag: a mobile game

This concept was developed in March 2002, in the light of already existing academic research in interactive applications that blur the physical with the virtual space, referenced in the literature review in chapter 2. The opportunity was evident: urban areas can get new meanings through location based play and at the same time social interaction can move beyond the computer and into the city streets. Commercial location-based games being developed in Scandinavia and in the UK focused on 'treasure hunt', 'hide-and-seek', clue collection and fighting. Because of technological limitations (WAP phones, only cell based location, small screens and limited input), we conceptualised a simple game, based on the idea of playground 'tag'. In this well known children's game, play emerges at often unpredictable moments, when, for example, a group of children walk along the street together and one touches another saying 'you're it' and they then start chasing each other. 'Tag' or 'tig' illustrates, like other playground games, that highly engaging playful interaction is possible even without the specific principles and goals of structured games.

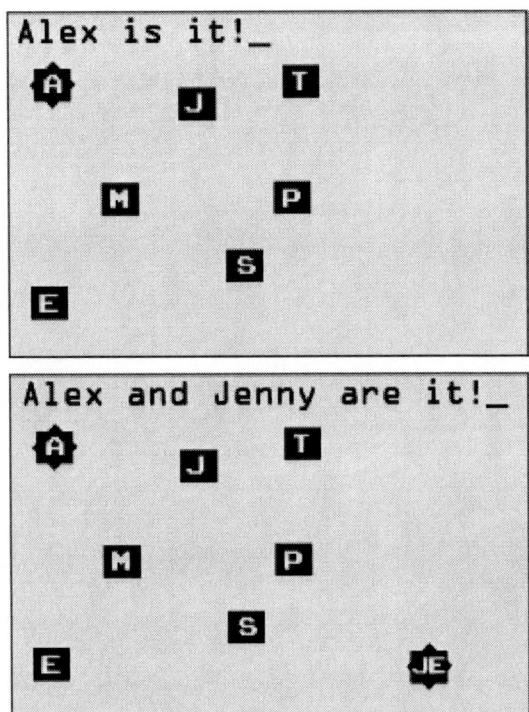

Figure 4.3 The single-chaser and two-chasers mock up screens of the game.

The Pixeltag game was a means to identify design guidelines for presence-based play on mobile phones. Pixeltag is a multiplayer chasing game on a mobile phone, which could be launched from an IM or chat application for phones by inviting people to play. We aimed to create a simple to understand, anytime, anyplace experience, targeting a broad and casual audience that would be looking for some quick fun to share with friends or other people during commuter time. One player is the chaser and tries to move to an adjacent to another player square to tag them (figure 4.3 a). Players move around towards any direction (up, down, left, right) by using the device arrow keys (or other 4 keys depending on the mobile phone design). Alternatively, in a turn-based scenario, players send an SMS message with their location on a grid structure (see figure 4.4 b), for example 1C. Every player would make a move and send it to the server, which would update each player's position and state (e.g. who is 'it'). The game is presence-based; if a player stopped playing their pixel would fade away on the other players' screens.

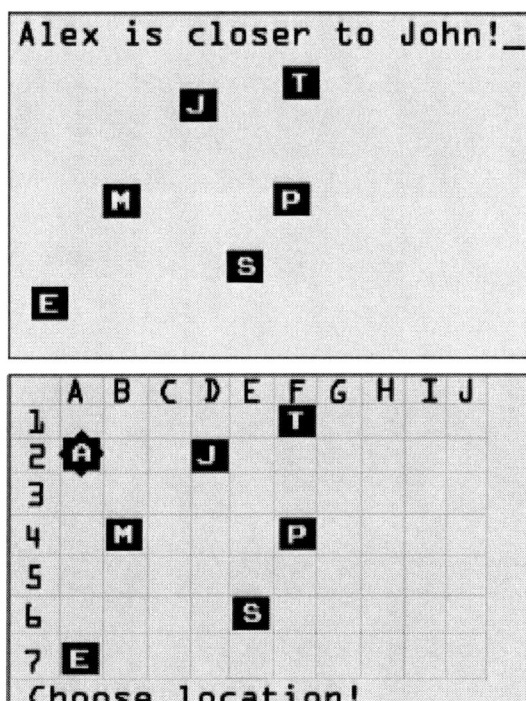

Figure 4.4 a) Here the hunter is invisible. b) Turn-based version: as a 'chaser' you can move to any direction, upwards, downwards, right or left as well as diagonal. The rest of the players can move upwards, downwards, right or left, but only 2 cells at a time.

In respect to our research framework in 2.2.5, the Pixeltag concept brought forward some issues on *communication, spontaneity, challenge* and *emergence*:

- *Communication and spontaneity.* What we found attractive about this concept is that it is simple to understand (children's playground game) and purely 'presence' based. Inviting players through an IM buddylist, was also appealing because it could bridge the two worlds of play and social software. Starting the game could be as spontaneous as the original 'tag', promoting a casual form of play not only among youngsters. The communication of the game itself is quite minimal, but the limitation of the screen size would not allow the current design to scale up representing more than 10 or so users.
- *Challenge.* The turn based version could accommodate different player styles (e.g. players who merely prefer to interact once per hour, day etc), also being within the limitations of mobile technologies for real-time data services (i.e. delays in response etc). We could also introduce variations to increase the levels of challenge: the 'hunter' would become invisible for a short period of time (figure 4.4a) and the other players would have to guess his/her move. Or we could have two competing 'hunters' instead of one (4.3 b). Other parameters such as varying the size of 'hunter' or player speed (i.e. how many 'pixels' players move at a time) could have different effects too. However, contrary to the challenges for the Maze game which require collaboration and aim to encourage emergent flocking, it is difficult to see how these challenges

could support collective behaviour. The only form of *collaboration* we might expect is that in the two-chaser version of the game the two chasers might try and move their pixels more strategically to trap the other players.

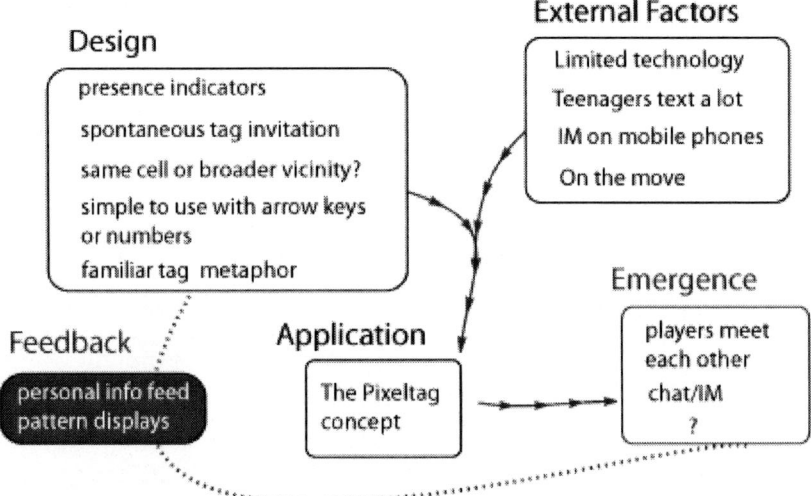

Figure 4.5 Speculative design for emergence with Pixeltag

- *Emergence*. Considering the kind of interactions that could emerge through this, we would expect that players' performance in the game could act as a springboard for conversation, either face to face or through an IM interface and they could possibly like to meet up in a physical location if they happened to be nearby when playing. In the design for emergence model for Pixeltag (figure 4.5), the main features include presence indicators, some broad physical proximity and the tag metaphor with a simply designed interface. External factors like texting habits, being on the move and the use of IM on mobile phones can influence the emergent interactions. There is yet no direct link with the physical world through the game itself, but we can speculate that players belonging to the same social group (e.g. classmates, 'buddies') are likely to meet in the real world and talk about the game, maybe even start a physical game of tag. Imagine that players used Pixeltag in completely unexpected ways, for example to create patterns on their mobile phone screen by positioning their pixels in a certain way. Again, we would only be able to find out by implementing the game and trying it with users. We need to consider what would encourage players to do that — a suggestive element in the design, for example being able to save game states as mobile phone screen savers. This could inspire the design of collaborative pattern creation: the user-generated patterns could appear in public displays or even buildings, similar to the concept of the existing Blinkenlights Project (2002), where an entire building was used as a computer display for animations and flirtatious picture messages and patterns. In our speculative emergence model (figure 4.5), the feedback in the design process suggests strengthening the social aspect of the game: players receiving information about the people they tagged, or some kind of personal trophy, like the tagged players' picture in pixels.

One problem with this concept is however that it represents a quite direct translation of the 'tag' concept into a primarily virtual interaction – there is no engagement with the physical world or the context of the location in which the interaction takes place. The Pixeltag concept is not very informative in relation to the *interaction models* in our research framework and the parameter of *behaviour in public*.

4.3 Mixed Reality games

The concepts in this section explore the design space for social interaction and collective play through ubiquitous computing technology, aiming to encourage spontaneous, emergent behaviours. Revisiting our design for emergence framework in chapter 3, we are interested in exploring the areas in the diagram in figure 4.6. What kind of experiment can we organize, based on a combination of ubiquitous computing design and external factors, that will enable us to observe emergent behaviours and unpredictable social uses of our original design? Can emergence inspire innovation? How would emergence affect our awareness as designers?

Following from Pixeltag, we considered other, more ubiquitous ideas for location-based multiplayer games that would actually take advantage of the players' physical and social environment and mobility, encouraging emergent interactions. The first of these was the TimeTravel game, a collaborative puzzle solving game which would have an educational as well as social value. The TimeTravel game would be developed for a PDA, third generation mobile phone or other device with location sensing technology (e.g. GPS) and enhanced graphic display. It is a puzzle solving game where players are trying to find a location on a hundred years old map of the area they are actually moving in. The old city map corresponds to a real city area of, for instance 1km radius, but displays different objects based on a fantasy theme. The only thing that is actually the same in the real and the virtual city is the presence of other people. One of the clues is crowd flow, as represented on this old map on the device (figure 4.7). This would indicate the busiest streets which a player would need to navigate. So if a player finds a crowded place he/she could guess the corresponding one on the map. As a player, you belong to a particular colour coded group that spreads anywhere around the city and you play against other groups. You try to guess your way around in the city and solve the puzzles, by observing the crowd flow and exchanging messages with your friends in other parts of the city. Your aim is to conquer parts of the fictional city as the narrative evolves in real time and the way to do this is to manage to go to a particular spot together with many other players of the same colour and send a message with your mobile phone. Depending on the number of people from your group that do the same thing at the same time you either conquer the place or not. If another group outnumbers you, the place 'belongs' to them. Different fictional places that correspond to the real in the city have different values (more or less points for conquering them) and people play the game all around the city. The challenge is to find which 'spots' get more points and go for these, at the same time making sure that your group has a significant presence there. By using the messaging facility players exchange information about clues or arrange to meet.

The concept relies on intensive *communication* (exchanging messages and clues), but at the same time it is also presence-based, since it encourages players to observe the presence of other people in a neighbourhood in order to make sense of the map. An appeal of this game concept is that it takes advantage of mobile gaming, messaging and location-based technology. Whereas the game takes place in the virtual realm, it links well to particular locations and we expect emergent interaction to happen in the physical world when people gather to conquer areas of the city. *Emergence* can be both collective (e.g. crowds gathering) as well as individual (e.g. meeting people like in Flash Mobs). The concept offers interesting dynamics on the parameter of *group belongingness* in our research framework: players are members of their own team but also belong to a larger crowd of people who participate in the game. The key issue in this game however is achieving a *critical mass*. If there are not enough people, players would not be able to observe differences in crowd flow on their maps to identify locations nor conquer parts of the city. This game would evolve in real time and require large scale participation. The TimeTravel game was very ambitious to implement as part of a research project, so we had come up with a simpler and more feasible concept, but we can expect to see some of these ideas in future location-based games. Clustering and swarming in the city as a form of *collaboration* and the overlay of the 'virtual past' on the 'physical present' provide inspiration for further ideas on mixed reality experiences.

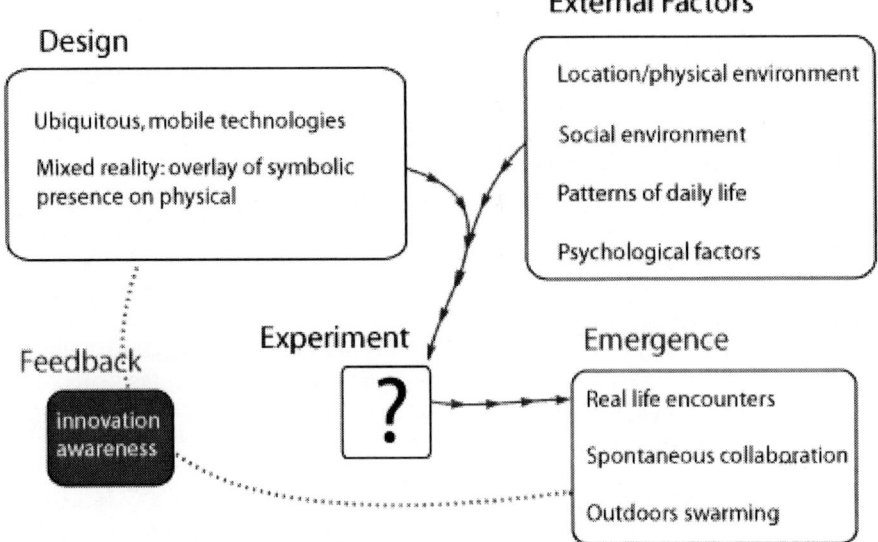

Figure 4.6 Design for emergence using ubiquitous computing technologies for spontaneous social interaction

At a later stage of the research, we developed several more scenarios on social mixed reality experiences. The following concepts can be taken further in future work and they are all part of the same theme: participating in a party with surprises and secret collective activities. The party would take place in an unknown location and people would not know who would be going there. Information about the event would be distributed with various media, email, flyers and a website. Anyone could participate provided they send a picture of a favourite part of their body or their clothing beforehand. In this way, participants must give away a little piece of their privacy to participate in a challenging event. On the day of the party, participants would be led through the city to the secret location via messages they would receive on their phones, and see in shop windows, on computer screens. These public screens would be waypoints guiding people to the location, but also displaying how many people already passed the screen on their way to the party, showing edited pictures. The screens could also spread rumours about those people and crazy activities happening (e.g. 12 people passed on their way to hugging). The idea is that people would start wondering who else is part of the party and that they would become visible to other party-goers.

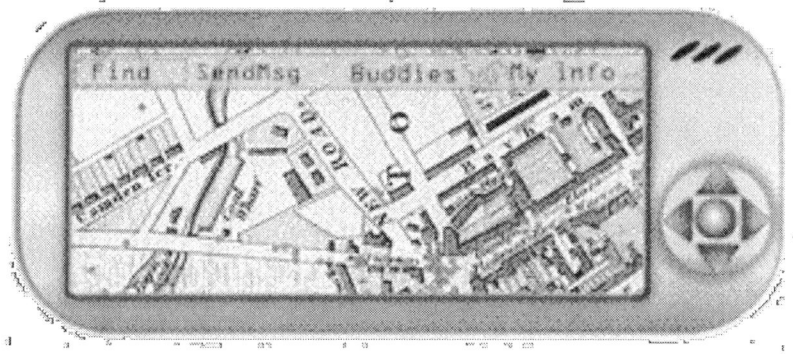

Figure 4.7 TimeTravel game: uncovering the old city by following crowd flows

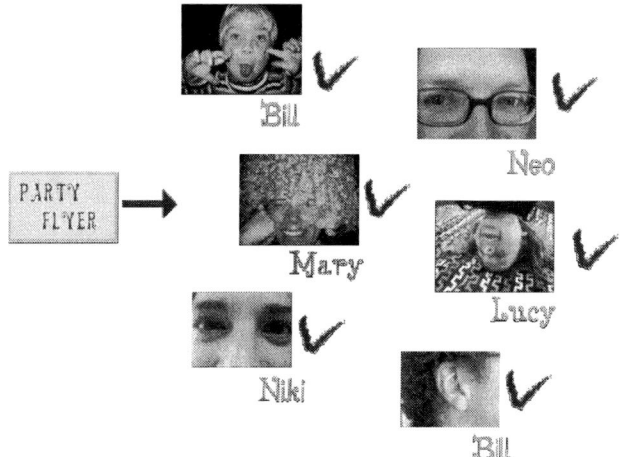

Figure 4.8 People need to upload or send pictures of themselves or part of their body or clothing in order to participate in the secret party.

At the party location, which could be a restaurant or some kind of club, a visual artist who would edit the pictures previously sent by the participants and would add more visual materials, displayed on a large screen. We thought of different families of games that could be played in this context as described below:

a) Group activities.

Participants try to find an interesting or favourite object in the location and take a picture of it, then to send the picture to the public billboard as part of a competition. Images can be accompanied with a short message, description or legend explaining the choice. Pictures are then grouped or matched according to similarity or some other criterion. So the visual artist can group these photos to create 'watch-people', 'shoe-people' and 'ring-people' etc. People are then challenged to find their pairs, to identify the people who took pictures similar to theirs by following clues sent to their personal devices and interacting with others around them. Participants also vote on the items, for example for the most unusual or funny picture, through their personal device. The results are then displayed on the public screen.

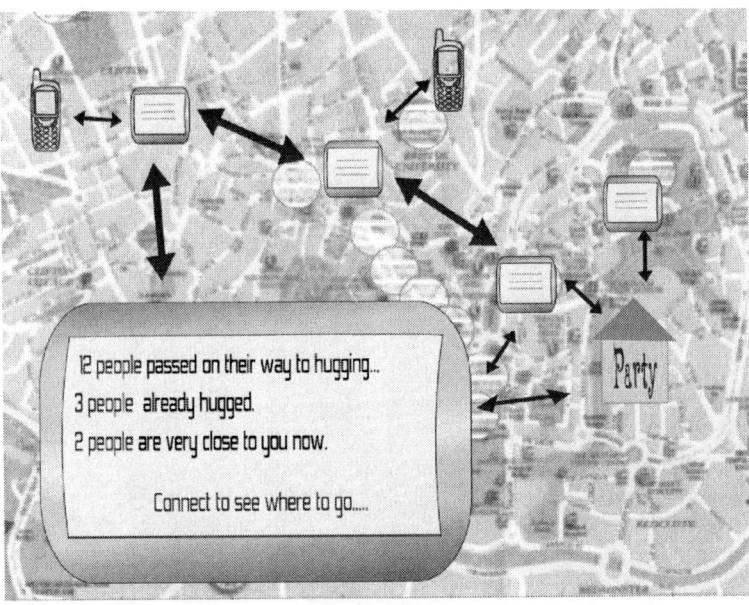

Figure 4.9 People receive clues and rumours on their way to the party location displayed on public screens and billboards.

b) Ambience – collective mood.

Participants' mood could affect a visual display or some other aspect of the environment: e.g. the lighting. Their personal device would include a 'moodometer' with 'melancholic', for example, displayed with dark colours and 'excited' displayed with bright colours. Moods can be: melancholic, sad, bored, relaxed, laid back, happy, social, adventurous, excited, humorous etc. Each person chooses his or her mood and the collective mood is reflected in the colour of the public screen or the lighting of the environment. Broadly, the colours in the red area of the colour spectrum have been referred as 'warm' and those in the blue and green range as 'cool' colours. Faber Birren (1978) has associated those with two moods: the warm colours, such as a red and its neighbouring hues, with excitement and the cool colours with feelings of calmness (e.g. blue, violet and green). Though these concepts are largely empirical, warmth and coolness in colour are dynamic qualities, warmth signifying contact with environment, coolness signifying withdrawal into oneself (Sasaki, 1991). In our collective mood display, more melancholic moods could give a purplish radiant, while excitement could be reflected with bright, reddish colours.

Another version of this could be the 'Mood Colour Wheel'. An animated colour wheel displayed on the public screen would comprise of colour blends/moods from each individual device moodometer. So each person would be represented by a part of the colour palette and each mood change would be dynamically reflected on the 'Mood Colour Wheel'. It would be interesting to interact with people of a particular mood in vicinity, for example to cause their device to vibrate or play a tune, or send them a funny picture. This could be the start of a new conversation or an emergent behaviour (e.g. many devices playing the same tune at the same time). The 'moodometer' running on each person's device could also control the music, resulting in a kind of jam session, where each device or group of devices with the same mood would control an audio track from an available selection of several tracks/loops.

c) Playing with 'Who is who?'

Another idea was to design a mystery, a puzzle type of game in which people would try to identify others by given clues, for example, in the context of the 'favourite object' poll mentioned above. The large screen would be displaying rumours about people who found or missed each other and players would be able to set alerts on their devices to notify them when a certain person is very close. Players would also interact with each other and cause funny things on others' devices: play a tune, vibrate, display a photo, a message, animation etc. Rather than a dating service, this would be a game, a fun experience to be shared with friends, also an opportunity to meet other people. Every now and then a puzzle would be displayed on the large screen, a photograph covered with square tiles with a question mark. Every time a player succeeds in finding another player, one of the covering squares would disappear, revealing the picture underneath.

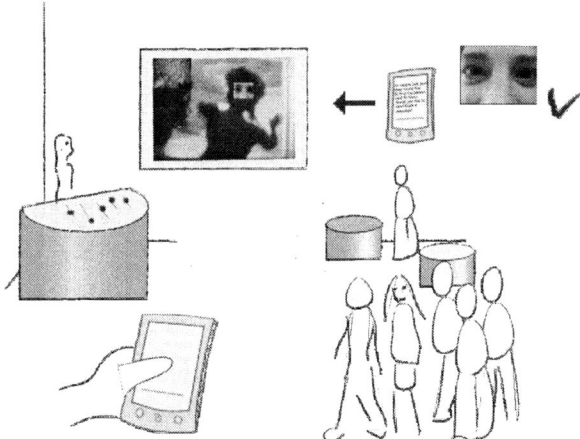

Figure 4.10 At the party a visual artist mixes the uploaded images and displays content received from participants' devices.

d) Bizarre Polls

Finally, we considered the idea of having funny polls, for example people voting on subjects like who is the best barman, waitress, the person with the largest 'beer belly' at the party, which one is the best drink etc. We envisaged this as being an interactive poll: participants would send their votes with their devices but they would also be able to introduce new polls.

All these ideas were developed with *emergence* in the main focus, where emergence is *an unexpected individual or group behaviour expressed in the real world, based or inspired by interaction through a virtual artefact.*

Figure 4.11 People vote on party related polls and participate in strange competitions involving funny activities, such as measuring aspects of others.

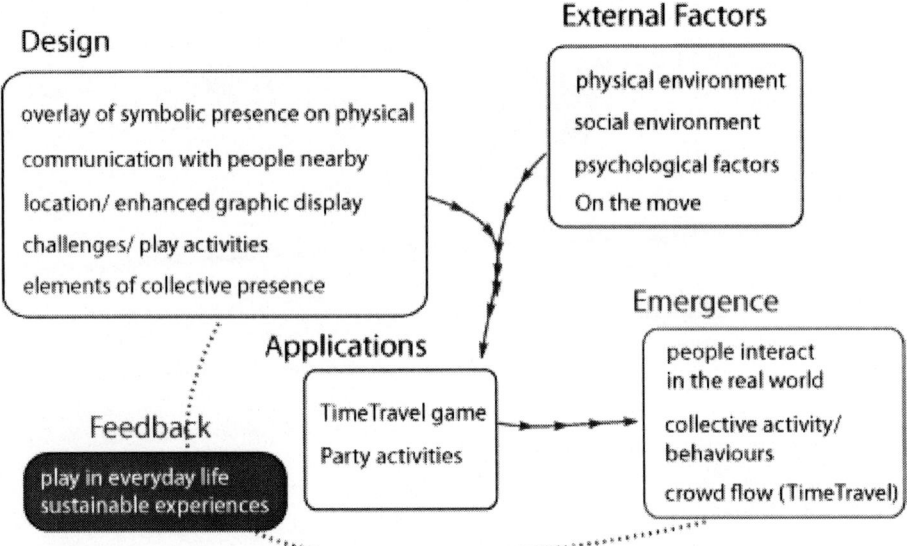

Figure 4.12 Design for emergence in mixed reality games and playful activities

Our speculative model of design for emergence applies to all the mixed reality experiences we just described (figure 4.12). The relationship between the virtual and the real world is tightened through the overlay of symbolic presence (e.g. dots on maps representing crowd flow in the TimeTravel game, pictures of people and favourite objects, in the party activities) on physical presence (real people at real locations). Depending on physical and social factors, like the numbers of people participating, objects found in the physical space as well as motivational and experiential factors, the emergent interaction can vary. In the TimeTravel game scenario we would expect the game affecting the crowd flow in the real world as people would gather at particular city locations in order to claim them. In the party scenarios, we can envisage these activities to spark social interactions, conversations, spontaneous behaviours (e.g. making other people's device vibrate, all playing the same tune) and engage the audience in a collective activity, like trying to change the general 'mood' in the 'moodometer' or taking pictures of each other, responding to polls etc. Experiments of this kind would provide feedback to the design process about ways to integrate play and bonding social experiences in our everyday life.

Considering the *experiential factors and variability* in our research framework, we need to understand which of these mixed reality experiences can be sustainable and developed further and which are just one-off interactions, spontaneous and engaging for participants, but not complex or interesting enough to be performed more than once. For example, in more participatory activities like finding 'who is who' or creating fun polls, the user experience would vary more than in the collective 'mood' display. The shared expressions of collective mood with the 'moodometer' and the 'Mood Colour Wheel' have greater potential to strengthen the sense of *group belongingness*, whereas the other activities are more individual. This would be a passive form of *group belongingness*, based on the manifestation of a presence state; people's mood. Responding to funny polls or sending a personal picture to the large display would indicate more active forms of being part of a group. These concepts do not enforce or require *collaboration* in the same way the previous do (e.g. cluster to open the exit in the Maze game or swarm the city to conquer areas in the TimeTravel game), however we would expect that participants would collaborate or at the very least communicate with other people at the party location when performing these playful activities.

The party activities encourage the expression of social *behaviour in a public space*. Some of these acts can be more embarrassing than others, depending on the extent to which the action is exposed or mediated (e.g. measuring someone's belly or approaching a stranger as opposed to secretly making someone's device vibrate). The sense of embarrassment would be influenced by the parameter of *group belongingness*, if people feel they are part of 'their own crowd', they are more likely to experiment more with daring activities. The public display also provides a good awareness mechanism for what is happening at the location. The *interaction model* in these activities suggests a flow of information between people's personal devices and the public display. The virtual actions are likely to be counterbalanced by real world activity, such as taking pictures of people nearby, discussing with friends about what to vote and so on.

4.4 Final thoughts towards a playground social game

The game concepts and scenarios described in this chapter worked as a starting point to develop ideas further. Most inspiring, reappearing in design concepts in this thesis, is the *spontaneity* of 'playground tag'. Our first attempt to bring this kind of interaction in the virtual realm appears in the Pixeltag concept above. However, Pixeltag was not taken further because it is somehow limited by the available technology; not taking full advantage of mobility and the context of a user's surrounding physical and social environment. Our aspiration to design a mixed reality social experience using ubiquitous technologies to explore emergence in the real world, led to the creation and experimentation with CitiTag, extensively discussed in chapter 6.

Drawing on our insights from the above concepts, we designed the BumperCar online game, which is discussed in detail in the next chapter. We envisage that the 'ideal' presence-based game: 1) has an element of *spontaneity*, close to our ideal of goal-less playground 'tag', like in Pixeltag 2) poses *challenges* like the Maze game 3) encourages both *planned* and *unplanned collaboration* through play (like in the Maze-TimeTravel and the party activities respectively) and can therefore facilitate *emergence* (e.g. crowd flow and swarms to conquer the city in TimeTravel) 4) enhances a sense of *group belongingness* (like the collective mood display) and 5) has a simpler and less structured design that the Maze game to allow for different playful experiences and unpredictable user behaviours – in other words *variability* of the user experience. In accordance to our principles outlined in Chapter 3, the game should also somehow relate to a real world and provide affordances for people to make sense of the environment. Following upon our principle of lightweight design, we need to provide 'just enough' context to see what kind of behaviours can emerge. We decided that the Maze game provides already quite a lot of context and although it is a promising idea to explore, we need to start from the simplest application possible, so that *complexity comes out of simplicity*, as we gain knowledge through the process of iterative design and user trials.

Part II

Online Case Study: Experiments with a Multiplayer Bumper Car Game

5. Playground Interaction Online: The Bumper Car Game

5.1 The idea

Our initial conceptual explorations in large-scale multiplayer games based on presence for emergent group behaviours and collaboration led us to develop the BumperCar game. This game appeared most appealing because of its simplicity and loose structure. We wanted to recreate the feeling of physical, playground games (Opie and Opie, 1969) and the engagement of ice-breaking party and group games that have no serious game goal, but are fun in their own right. In a similar spirit to 'tag', we designed and developed the online Bumper Car environment described in this chapter, as a 'playground space' to be used spontaneously for different playful social activities. The game could potentially be used as part of an Instant Messaging environment, like Buddyspace – a communication and collaboration tool incorporating advanced presence awareness features for Open University students (Vogiazou et al, 2005). BuddySpace generalizes the concept of 'buddy list' to include optional geo-location and personal profile data – this enables a mixture of personal and automatically-generated contacts to be displayed with their real-time presence information in a custom and compelling visualisation (Eisenstadt & Komzak, 2005). With the BumperCar game we aimed to bridge *the immediacy of online symbolic presence as in Instant Messaging environments* with the *unpredictability and engagement of spontaneous, playground-like social play*. We also envisioned the Bumper Car game to be used as part of other computer supported collaboration and learning tools, as a means to develop a community and to facilitate informal encounters between isolated individuals online.

We aimed to find out in what ways people's presence can be used to facilitate chance encounters, collective recreational activities and ad hoc social interaction, based on the awareness of 'what other people are doing at the moment'. Non-verbal communications, spontaneous interactions, informal and physical presence are all elements of face-to-face interaction that can promote a sense of community. In real world playground tag, one child challenges another by touching him or her and saying 'You're it!' and this is how a spontaneous playful activity starts. With this interaction in mind we imagined IM users would invite their 'buddies' to play a quick game, during a break from work, as a stress relief, or just to get the opportunity to get in touch with others.

Some of the broader research questions we address with the Bumper Car project are:

- How can we enhance social interaction and spontaneous collective behaviours online through play among large numbers of people?
- What are the affordances we need to provide and how can online participants relate to the virtual environment and extend it?
- Can visual information in the form of symbolic presence 'states' be sufficient for participants to understand what is happening and maintain awareness of what other people are doing?
- What kind of cooperation challenges can we introduce in the environment to vary the player experience and encourage coordination practices?

- What are the physical (movement, collision), visual (ability to see and interact with other players) and psychological (fun, engagement, feeling social) constraints in this game?

Our 'playground space' is founded on the following four design principles detailed in chapter 3:

1. *Presence is largely symbolic:* with a simple, symbolic visual design, we enable abstract communication, such as changing the colour 'state' to convey group membership.
2. *Large scale is important for emergent interaction*: players drive cars that can be scaled down to small dots as the number of participants increases.
3. *Keep the design lightweight*: by creating a simple, cartoon style, 2-D environment we cater for scalability. The 'organic' look and feel of the game was inspired by gas and physics educational online simulations. At a large scale, these paradigms can help to make the presence of large numbers of players perceivable and visualize the flow of online 'crowds'.
4. *Employ affordances*: our environment is designed to accommodate emergent social behaviours, much like 'flocks' or self-organized groups, by encouraging people to use their presence in playful and creative ways. The metaphor of BumperCars is appropriate, because it easily relates to a familiar, physical real world experience in amusement parks. There is something inherently 'fun' in bumping into each other's cars in the real world arena, and we believe it is possible to translate this into goal-less participatory play online. The concept also encourages an exploratory approach, to try and experiment with the physics of bumping and moving around. We also employ a visualisation of the entire virtual space, so that everyone can see the overall world gestalt.

The Bumper Car game prototype we have been experimenting with, is our first attempt to design for emergence by deploying a minimal design, just enough of an environment and context in order to see what kind of behaviours will emerge (if any). The design process of the Bumper Car game was far from straightforward because there were too many unknowns in terms of both the final outcome that we were trying to achieve as well as the way to achieve it. In the following paragraph, we discuss the evolution of the idea and the different variations we considered in order to trigger some kind of emergence, a collective behaviour through play.

5.2   Storyboards and variations

The Bumper Car game went through several conceptual phases before we concluded with the final prototype. Some of the first questions we asked were 'What will happen in a massive game with engaging physical interaction but with no goal? How would people behave? Would we have any emerging behaviours?'

Conceptualising a goal-less game was one of the most difficult design challenges in this thesis. Drawing from the game design world, we know that there is always an element of challenge, associated with a goal or outcome (Salen and Zimmerman, 2004). How would people relate to our 'playground space'? We wanted to avoid the pitfall of designing an environment where users would not know what to do, driving and bumping around a bit and then leaving without having enjoyed it.

Table 5.1 maps game variations with a collaborative or a competitive aspect, encouraging team based play. On the vertical column we have three variations based on the team/group parameter: the first version has no groups at all, the second has group divisions based on plain colour coding while the third version would be more structured and incorporate a group formation mechanism (flocking of three or more cars to form a powerful group). On the horizontal axis we have two variations, a collaborative one, where players gain strength and points by cooperating with their teams and swarming and a competitive version, in which individual players or teams play against other players or 'flocks'. The latter is based on the 'survival of the fittest' concept and it is closer to the notion of typical goal-oriented games than playground, unstructured play. We can also include additional challenges, like the 'exit' challenge, based on the idea of the Mintz experiment we discussed in chapter 2, posing various dilemmas.

Elements like the group identification based on colour coding and 'flocking' to become stronger aim to extend the simple bumper car concept towards collaborative group play. Such collective behaviours would distinguish a participatory presence-based game from existing genres of online multiplayer games.

The formation of self-organized teams raised some important issues. On the one hand, it could provide opportunities for interesting play and emergent group behaviours. We would expect to observe swarms and group strategies evolve over time. On the other hand, decision-making on the direction of movement and individual influences on the overall group performance posed challenging trade-offs. If a leader decides which way the group will go, what would be the role of the other group members and how would they maintain engagement? We considered voting mechanisms, assigning different roles, including messaging facilities, but all this would add complexity to the game and the interface – working against our principle of lightweight design for 'just enough' context to allow players to discover such balances through experimentation.

Table 5.1 Bumper Car variations based on collaborative, competitive and team aspects

Group division and colour coding	Collaborative	Competitive (offensive)	Additional challenges: the 'Exit'
1. No teams, everyone has the same color and size	As a player you lose energy (fade) when bumping other cars. Therefore you try to avoid offensive players and be defensive. This is a starting point for other versions.	Simple Bumper Car Game (A testbed for physics and collision)	Whoever gets faster out of the exit gets a higher score. Players also gain scores for avoiding collision on the way, so they are facing the trade-off between driving fast and giving way to avoid collision.

2.Colour coded groups. We can have at least 2 teams or more.	We will see if any group behaviours emerge just by having cars different in colour and size. Example ideas: if you avoid hits you grow in size or if you get bumped too much you loose size. Will people behave differently to their own colour group?	Collision against the opposite group members makes you gain energy (brighten), but collision against cars of the same colour make you loose energy (fade), e.g green-red=gain energy, get brighter green-green= fade a bit	You are too big to get out of the exit-bump against the opposite colour group to become smaller. But then you move very slowly. Bump against your own colour group to gain speed (Alice in Wonderland effect). Interesting to see what would happen.
3.Self-organised teams. Colour coded groups are formed with a group formation mechanism. One player chooses colour from an 8 colour palette and invites other players to join the group.	You gain strength by forming groups. You start playing with a faded, pale colour and by forming groups you gain strength and become brighter (or grow in size).	You create 'squads' of 3 cars or so and this gives you huge group power, you can beat (e.g. fade out) a single car with one bump. Either one 'leader' drives the 'squad' or they practice to drive together (group-gravity effect).	The exit will open only when one team achieves a large size, a certain minimum number of members (e.g. 10 players). Players break up from their teams and run for the exit.

Using visual communication seemed more appropriate to facilitate ad hoc coordination, for example a visual cue (e.g. an arrow) to indicate each player's direction of movement. Similarly, colour would suggest team membership, so if a player wanted to leave their team, they would change their colour to imply a different identity. Colour would also work as an indicator to identify possible team mates to link with (same colour) or players to bump into (different colour).

Additional challenges, like the 'exit' in table 5.1 following from the Mintz experiment described in chapter 2 would add interesting social dimensions to the game: a trade-off between individualistic play (e.g. run for the exit) and cooperation (e.g. negotiate how you get out to avoid collisions). Before describing the design of the final prototype, we outline some trade-offs and technological issues we had to consider.

5.3   Technical limitations and design considerations

One of the first issues we encountered when designing the game was the kind of inter-player communication channel we should provide. Messaging and chat would pose an additional cognitive overload as players would need to pause their game activity in order to type a message. Then we considered the use of predefined messages, like in first-person-shooter games. However, this was problematic as well, because the first question that comes to mind is: what would people want to communicate in a free play environment that does not involve a very specific goal like 'capturing the flag'? Moreover, text messaging is not a scalable communication tool and neither is voice – another alternative. Voice communication would strain the network and slow down the game server as well. So, we decided to implement a prototype based on visual communication, without text messaging or voice.

At first we considered implementing the game in Macromedia Flash MX to take advantage of its design flexibility and create the organic, fluid look and feel we wanted, while keeping it lightweight by using a vector based program. However, the Flash Communication Server which was just rolled out by Macromedia at the time (spring 2002) could not support real-time interactions among many simultaneous users very well, so the lag for over eight to ten participants would be quite significant. We decided to implement the game in the Java programming language, which appeared to be more reliable in terms of server-client communication and managing the connections between multiple clients.

We did experience problems and technical limitations with Java too. The programmer who implemented the game software found that if at least one client ran on a slower computer this would slow down all communications. Also, if more than twelve people were connected at a time the game would slow down too much. Therefore we had to carry out the study within those limitations, given the time and resource constraints.

In order to run a series of experiments with distributed participants across the Open University campus, we needed to be able to communicate with them simultaneously during the game in case something went wrong, for example if they had trouble connecting to the game, if they were experiencing lag or even just to signal the start and end of each game experiment. For this reason we decided to run a separate chat facility alongside the game, in order to coordinate the start and end of the experiment and to be able to help people with problems on a separate communication channel, independently of the game software and server. With this facility we could also have a chat break during the trial, if necessary for people to discuss strategies and make suggestions about their performance in the game. The chat program was written in Macromedia Flash and was therefore easily accessible from any web browser with the Flash plug-in. Our Bumper Car game client required the installation of the Java Runtime Environment 1.4 before running the game.

We describe the design of the final Bumper Car prototype next.

5.4 BumperCar Design and Experiments

5.4.1. The game design

The Bumper Car game was written in Java, using custom made graphics to provide the desired look and feel. It consists of a multi-user server, with an administrative interface to create variable games and a downloadable client, to be installed on the user's personal computer as an executable file.

The server's administrative interface is used to specify various parameters of the game and manage different games. When creating a Bumper Car game, the administrator can vary game duration, player speed, change image background and add free-floating moving cars ('bots') to create different games. By adding 'bots' we can recreate the atmosphere of many people playing, even if there were a small number of actual players. We used this feature to test the visual and physical limitations imposed by increasing the number of cars within the environment. Then we created appropriately sized game spaces (backgrounds) to accommodate more people. By changing backgrounds we can change the context of the game and create different playful activities as the ones in the following section 5.4.2.

Each player logs in the game by entering a user name and password of their choice in the game client. The client then connects to the server and the player is assigned a car, which appears in the 'playground' area. The cars can have four colours: orange, green, blue or purple. Players use the keyboard arrow keys to move their car, bump into other cars and to change the colour of their car by pressing an appropriate key (e.g. B for blue, O for orange etc). A player can see other cars with their player's name tag, drive around and bump into them.

Inspired by real bumper cars in amusement parks, where people tend to target a particular car (usually that of their friends) and chase it until they bump into it just for fun, we decided to include the idea of 'bumping intention'. So a player could declare 'I am going for that target car now'. Players can 'put someone on the spot' or place a 'challenge', by clicking on another car of a different colour. By placing a 'challenge' on a 'rival' car we immediately have a hunter and a runaway player, in other words a 'challenger' and a 'challenged' player. A line is then drawn between the challenger and the challenged car (figure 5.1), which stays for a limited time, depending on the initial distance of the two cars. Within this time limit, the player who initiated the 'challenge' must reach the other car and bump into it to gain points. Challenges cannot be placed between cars of the same colour, since they are considered allies. If a player changes colour during a 'challenge' to avoid his or her chaser this does not take effect at that point and they are still challenged. A visual element, a 'glow' appearing around the car rewards a successful challenge (for the 'challenger') or successful collision avoidance (for the 'challenged' player).

In order to communicate the presence of all people playing simultaneously, we provided the facility of a single overview map, where cars are scaled to small circles or dots (figure 5.2). Players can switch quickly between the two views (normal, close-up and map view) by pressing the 'M' key of the keyboard for map. This feature addresses the playability trade-off between (i) the benefit of being able to interact while observing the whole world gestalt, on the one hand, vs. (ii) the disadvantage of having less of the immediacy and salience of seeing one's own car and region in the close-up view (Vogiazou and Eisenstadt, 2003).

With this simple, symbolic visual design we encourage abstract, visual communication, such as changing the car's colour to indicate alliance. The aim is to give 'just enough' for people to use their presence in playful and creative ways.

We tested the prototype with a total of fifteen people (lab colleagues) with varied gaming experience. We run a couple of game sessions for about 20 minutes with twelve and eight players respectively and ten 'bot cars' moving and bouncing around continuously. We observed the gameplay and collected participant feedback through questionnaires and a group discussion after the first test with twelve participants.

People used the colour change facility in various ways, for example to trick other people in order to challenge them unexpectedly or to avoid revenge by players they had just challenged. This is an example of an emergent behaviour that was not part of the original game design.

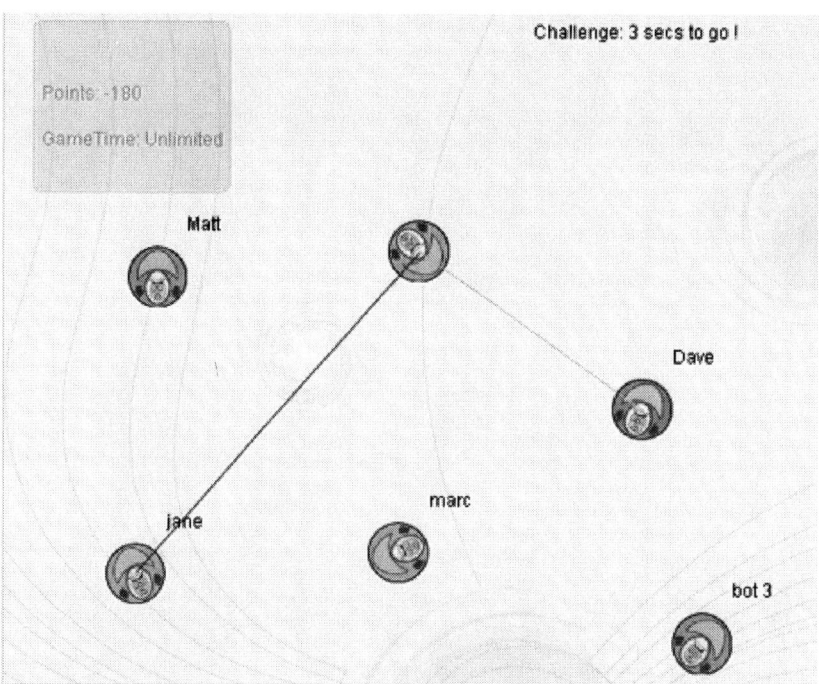

Figure 5.1 The user, whose screen we are observing, has challenged the player named Dave and has simultaneously been challenged by Jane. A green line indicates challenges posed by the user while a red line indicates challenges placed on the user by others. On the top left the user sees the remaining time within which the challenge must be completed (in this case he or she has to bump Dave within 3 seconds).

Twelve out of fifteen pilot participants gave positive comments about the game: 'instant fun', 'I liked the graphics simplicity', 'fast and addictive', 'easy to pick up'. One player in particular said: 'It's good fun to bump your boss or colleague!'. Nine out of fifteen people reported they would like to have an explicit way to team up. In particular, players suggested they wanted to 'chase other cars together' and 'form alliances against others'. Some players proposed ideas to enhance group identity and collaboration – for instance, 'knowing who else is being chased by team players'.

Eleven out of fifteen people reported they found the map view useful. Players used the map view to locate others, but then switched to normal view to place their challenge, since it was not possible to place challenges in the map view. Thus, it became apparent from users' feedback that full game functionality should be provided in both views. However, we know that at some point there has to be a limit to the 'playability' of very small bumper cars on very large maps. The key is to define this limit as we scale up the game, while bearing in mind the trade-off mentioned earlier between the experience of the whole world gestalt and the immediacy and salience of the perceptual view of one's own car.

This pilot trial of the Bumper Car game showed that there is good potential for engaging play and that people are keen to play collaboratively. In the next step of our research, we varied the game to encourage self-organization and to see whether people can coordinate as a group. Fascinated by the potential of emerging group behaviours, we wanted to design playful applications that people could use to surprise us with spontaneous, synchronised collective actions, such as changing colour in a particular way to achieve a rhythmic effect. In this work we addressed a particularly difficult trade-off: on the one hand we aimed to study group behaviour, motivation, collaboration and visual communication in a gaming context, but on the other hand we attempted to do this by creating a 'playground for emergent presence-based play', where normal channels of communication and stereotypical game-playing goals are restricted.

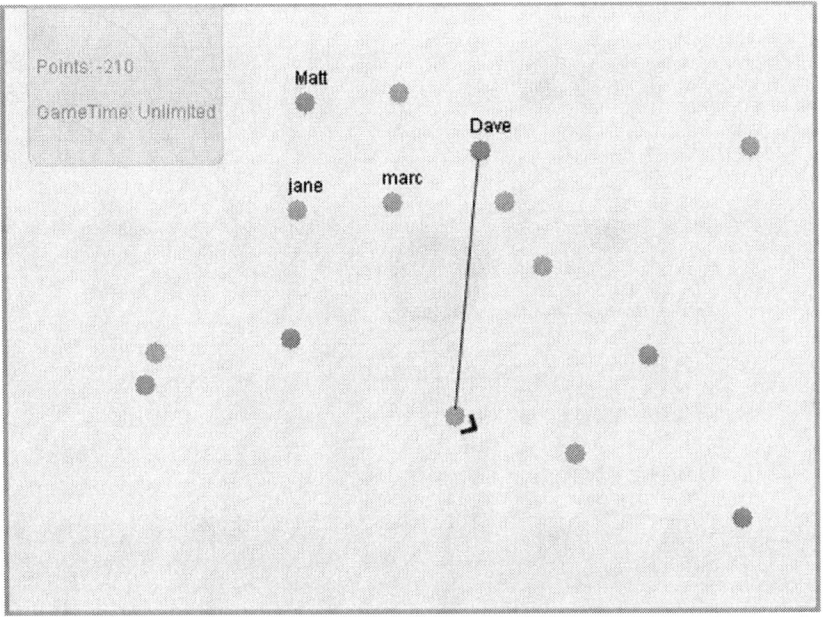

Figure 5.2 This is the map view of a user who has been challenged by Dave. The arrow indicates his or her direction of movement in this view. There are several free-floating bot-cars moving around (the coloured circles with no name).

5.4.2. Experimental Design

We designed a series of experiments and recruited participants through a university wide email list. Interested participants were invited to sign up a form on a web page, indicating their availability from a set of suggested dates for the online experiments. Details were finalised in a follow-up email. In total, twenty two people from various departments within the Open University campus participated, each one from his/her own office using a desktop computer. Coordinating a distributed experiment of this nature, with several simultaneous users at different locations was quite challenging. As mentioned in 5.3 above, we ran a separate chat facility alongside the game, in order to coordinate the start and end of the experiment and to be able to help people with problems on a separate communication channel. We used this only in the first couple of experiments and then replaced it with a set of visual instructions to coordinate the game sessions, for reasons we explain in detail further in section 5.5.4.

All participants were sent an email detailing system hardware and software requirements and instructions for the installation of the Java Runtime environment and the game software well in advance. They were also asked to login in the game and perform a technical test to ensure that everything was working before the experiment.

The six online experiments lasted about 20 minutes each and we recorded all onscreen interactions on video. At first we considered using a screen recording software, like Camstasia, but when testing it we found that it slowed down the graphics rendering on screen and the game, so we chose the more traditional method of a fixed camera view. We collected participant feedback by email questionnaires as well as through informal face to face follow-up discussions. Each experiment had 5-10 participants, plus one 'facilitator': the researcher logged in as an 'idle' bumper car to signal the start of and end of each game. All recordings were taken from the facilitator's screen. Many subjects participated in more than one game, since there were 3 different games to play. We collected 36 questionnaires in total and recorded approximately 2 hours of video.

For all the experiments we used the Bumper Car environment as a basis and we only changed the background image of the game, added bots (free-floating bumper cars) where necessary and suggested rules of play in the instructions we sent to participants, to create the three different playful online activities described in the following paragraphs. Our studies spanned contexts of interaction ranging from a less goal-oriented and unplanned activity (jam session) to goal-oriented play (collaborative pong), with the activity of group formations being somewhere between free play and goal-oriented games. Each type of the three activities was run over two separate sessions, yielding six experimental sessions.

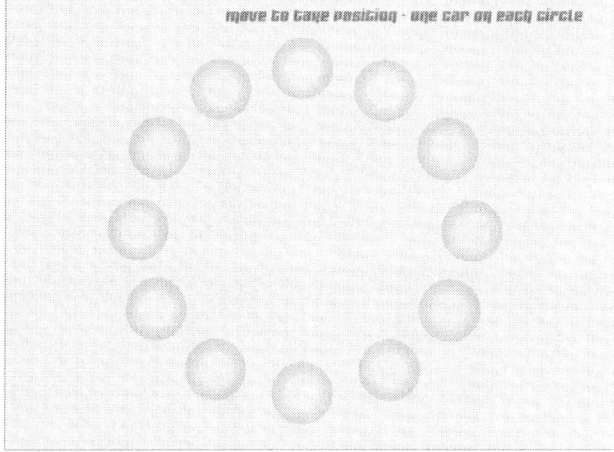

Figure 5.3. These were parking spaces for participants to use in the colour jam session to experiment with spinning and colour changing combinations.

a) Jam Session
In the first two experiments we created a *colour change jam session*. In the first experiment, participants were invited to try and create synchronised colour changes around a circle (figure 5.3) as if they were participating in a 'group jazz dance' competition, a kind of 'free expression' activity – though with abstract and deliberately loosely defined artistic criteria. Colour change and spinning were the suggested forms of expression. Eleven people participated in this first experiment. In the second experiment we introduced a team variant, making the game both collaborative and competitive at the same time. The eight participants in this session were asked to form two teams and to try to make rhythmic synchronised colour changes together with their team members. A separate chat facility was available in these experiments. In the

second experiment we introduced a set of visual signals from the facilitator (figure 5.4), which proved much more efficient for the coordination of the session than the chat. Therefore, the chat facility was removed from all the experiments that followed. As we see in figure 5.4 on top of the screen, there are 5 icons indicating different phases and instructions for the game session (start when I stop spinning, swap places, everybody turn orange, go to chat and experiment finished). The facilitator moved on top of a relevant icon to communicate an action within the game. For example, if the session was getting chaotic and there was a need to bring everyone to a starting point, the facilitator would move to the icon saying 'stop, everyone turn orange' in order to get everyone's attention and start again. The facilitator would then move on to 'start when I stop spinning' to spin around and stop, signaling a restart of the session.

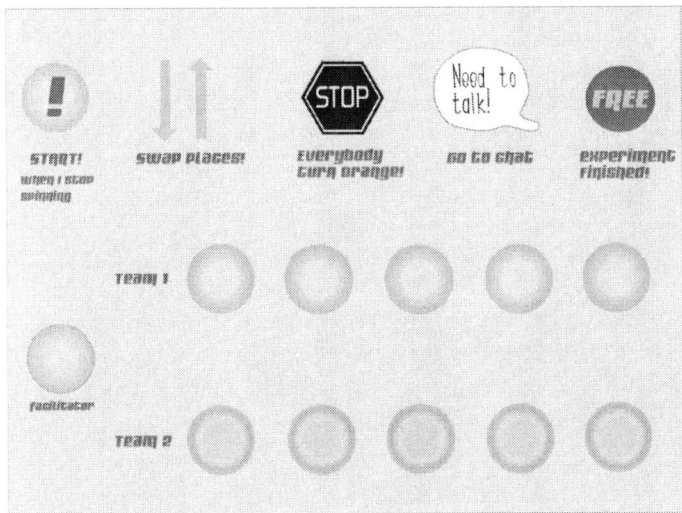

Figure 5.4 The team variant for the colour jam session with visual instructions on top.

Figure 5.5 The 'signal' for the start and end of the session was performed by the facilitator moving on one of these icons, each located at the bottom left and right corner of the game environment respectively.

b) Group Formations
In the experiment titled *group formations and chasing,* participants were asked to form groups on the fly, based on their colour identity and then to try to chase individuals of a different colour. Players could also change their own colour depending on what was happening in the game and whom they wanted to team up with at any time by pressing

the appropriate keyboard keys. We did not implement any game-specific group formation mechanism, because a) we wanted to see to what extent players could perform collaborative movement and chases together ad hoc without a binding design that would impose a flocking behaviour and b) time and resource constraints did not allow us to make such major changes to the software that would require a lot of further development, testing and debugging work.

In group formations, the facilitator moved on one of two visual icons (figure 5.5), to signal the start and end of the game session. Five people participated in the first group formations experiment and eight in the second. We did not run a parallel chat facility. These were the only two out of six experiments in which the map view was used. The entire environment was a bit larger than the game screen to give some more space for movement. In the rest of the experiments the close-up/normal view was sufficient since we had a fixed space for each activity, so there was no need to use a larger overview map or provide additional space. Because we had small groups and subsequently small environments, only a relatively small part of the whole view was 'out of range' in the close-up view in these two group formations experiments. So, one could spend most time in the normal view, occasionally switching to the map to locate others. The map view eliminated some visual information (the direction of other cars) but provided information missing in the close-up view (all participants' location).

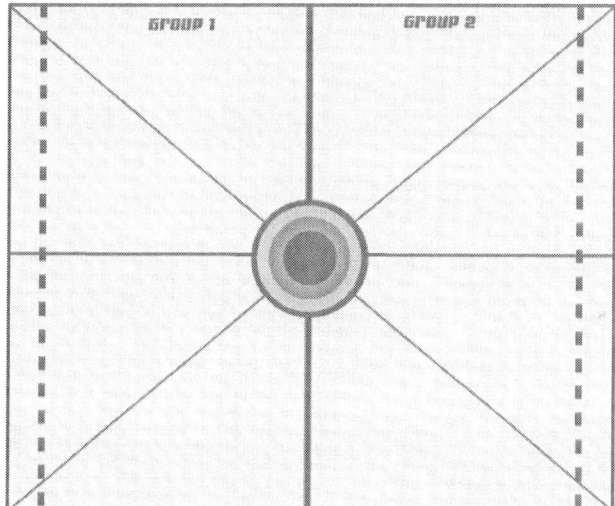

Figure 5.6 The background set up for the Pong game, a division of space among two teams, each one 'guarding' either side of the screen by bumping off a free-floating bot car.

c) Collaborative Pong

The collaborative Pong game was the most goal-oriented activity we organised, with the strongest team aspect. Participants were divided into two teams (green and orange) in advance to play a Pong game by defending either side of the screen (figure 5.6) and trying to send the ball (in fact, a free-floating bumper car 'bot') towards the opposite side. Teams were fixed for the whole duration of the session, so unlike the other experiments, a player could not join the other team by changing colour. This activity was closer to the concept of a structured game and with this we aimed to compare the participant experience in relation to the other two forms of play we tried. Seven people participated in the first Pong session we organised and five in the second one, with one however, not being able to really participate because of technical problems, so in reality the second game was played among four people with an additional participant just observing what was happening.

5.4.3. Analysis method

The findings reported in the following sections came out through qualitative research: analysis of video, two chat logs and responses to 36 questionnaires, as well as some comments several participants made in follow-up discussions. We tried to correlate interesting behaviours and events in those game sessions with participant comments, wherever available.

When analysing the video of the two colour jam experiments we used a set of codes, numbering and additional notes to record: a) who started a colour/spin (clockwise or anticlockwise) spread b) who followed by copying the pattern c) who changed the pattern d) any interesting, unusual emergent patterns e) synchronisation among members of one or both teams and f) who was idle or not participating. We noted changes and events for every 10 seconds in the video, adding numbered references to more detailed notes on what the facilitator was doing at the time, any relevant chat lines, breakdowns in synchronisation as well as particularly interesting behaviours and unpredictable events. The 10 second sampling was sufficient to get all significant changes in the state of the experiment. People changed 1-3 colours on average during a 10 second period (usually not more than 2, as they were also spinning at the same time). These were the codes used in the analysis, along with more detailed notes:

CN= Copy a neighbour (note whom).
'I'= idle, not making any movement
'O'= idle in orange colour, which was the starting state for everyone after the facilitator would go to the icon 'everyone turn orange'
CNW= Copy neighbour in a wave-like pattern, when a colour/spin spreads across a team. Participants were numbered (1,2,3 etc) in the order they changed their colour to keep the wave going.
COT= Copy opposite team member (this was less frequent).
CT= Change teams (when people did it spontaneously, not when instructed).
MB= Move around bouncing into other cars (spontaneous behaviour).
MP pattern = a pattern of moving around one's parking space in a circle – initiated by one participant and repeated many times in the game.

The analysis table can be described as follows: on the horizontal top line there are time stamps of 10 seconds and on the vertical axis the names of all participants, including one category for team 1 and one for team 2, for the second, team-based experiment only. Because teams were dynamic, changing many times during the game through 'place swapping', it was hard to evaluate every team's performance, since members were shifting to the other team frequently. However, these two categories were useful for noting interesting things that happened within one team as a whole, e.g. if all members had the same colour and/or spin direction for some time, or the same behaviour etc. We also included a last category 'teams in relation' to write all the events that happened to both teams, e.g. when one team was copying the colours of the other or when they were both doing the same thing.

This analysis helped us identify how synchronisation emerged, who were the most active individuals that took a 'leadership' or 'innovator' role by introducing new patterns for others to follow, how patterns became collective, what was easy and what was difficult for people to follow up upon and so on.

For the collaborative Pong experiments, the sampling rate for the video analysis was 20 seconds. In this case, we did not have any colour change but movement. We noted each participant's movement in relation to other participants. Individual behaviour was observed to see whether there were any emerging patterns, for example one participant being always in defense, or another always attacking the other team. Therefore, 20 seconds was a good sampling rate to note any interesting change in the overall position and behaviour of each participant and each team.

Teams were fixed, so we divided all participants in Team 1 and Team 2 on the vertical axis of the analysis table. Having read the questionnaire comments, the aim was to note signs of collaboration between team members, whenever possible. A clear sign was for example, when a player's move would make another team member move immediately to another direction to cover up an occurring gap. So these were the codes we used in this case:

D= defensive, when a participant stays back to defend his/her team's side whether moving from one spot to another or staying at the same spot.
    A= attacker, when a participant goes to the side of the opposite teams and tries to direct the ball to their border line.
    CHB = chase the ball. Many times this was just a chase, there didn't seem to be a clear attack or defense strategy.
    PB= Pass (bounce off) the ball to someone else
    DIV- U/L = Division of space between team members: upper/lower.
    DIV – F/B = Division of space between team members: front/back.
    FB = Form a block with one or more other team members (noting the exact number) to bounce the ball off their line.
    F= any other interesting type of formation that needs to be commented/ recorded in detail.

We often had combinations of codes in a 20 second slot, e.g. D with DIV – U/L. We also included numbered references on any other individual strategies/ positioning in space or other interesting observations, like breakdowns in collaboration. In this way we identified patterns of behaviour both at the individual and the team level.

Our general principle for sampling in this kind of experiments was a) to try and identify the behaviours and events that are relevant for the study and b) to note down their occurrence based on an approximate estimate of frequency. So for example, in the collaborative Pong game, the most interesting changes in how people moved around in space occurred every 15 seconds or so, thus a sampling of 20 seconds was sufficient. In the colour jam sessions, colour changes were more frequent, so we used a 10 second sampling.

For the group formations experiments, we used a less structured form of analysis, which was less time consuming, yet sufficient to record all relevant events. Again, we noted our observations with a time stamp, but without a regular sampling rate and the detailed table. We wrote a list of events, taking notes on the following:

a) Whenever participants made or responded to a colour flash signal to attract attention, including their reactions, such as whether they started to move along together or moved away etc.

b) All the collaborative chases of two or more people (noting participant names too). We noted every time people of the same colour (i.e. belonging to the same team) moved along together for a while in a chase and when they managed to trap a bot or another participant together (usually in corners).

c) How teams were formed in the first place and also which teams (2,3,4 member or more) seemed to work better, i.e. were more persistent and more coordinated in their moves and chases. We correlated participants' comments to our observations.

d) Spontaneous behaviour and expression. What participants did beyond instructions or the game context. Participants' comments about those in the questionnaire were useful as well.

e) Any evident individual strategies and behaviours, correlating those with participants' questionnaire responses.

5.5 Findings

We summarise our findings from the six experiments with the Bumper Car game below. Firstly we discuss emergence, unpredictable participant behaviours that occurred beyond the game context. Then we describe cases of group collaboration within the context of the specified playful activity and discuss influential relevant factors. We outline the main aspects of the players experience and our conclusions on visual communication and design.

5.5.1 Emergence

We observed several illuminating examples through our video analysis in which certain behaviours emerged spontaneously. This happened when people used aspects of the game design, such as colour changing in unexpected ways. For example, at the end of the second group formations experiment four participants who remained in the environment clustered spontaneously on one spot, almost one car on top of the other. What is particularly interesting is that nobody instructed this, it was not part of any guidance or game rules and there was no way to communicate this idea during the game, because the chat facility was not used. The way we can describe it is that when the experiment finished one person approached and tried to 'squeeze' the game facilitator and then the rest just followed. Everybody ended being on top of one other, like in a big 'group hug' (figure 5.7).

In the first group formations session there were two participants who came up with a spontaneous 'victory celebration' dance pattern (as one of them described it in a follow-up discussion) – they rotated in a lively manner around themselves every time they succeeded bumping someone else together. They managed to synchronise nicely without this being part of the game instructions or having any purpose at all. Again, there was no verbal communication, so these unplanned collaborative behaviours emerged in an impromptu fashion. This underscores our principle of symbolic presence; we can see that, when two people paid close attention to each other in a given context, a certain movement acquired a new meaning (dancing to celebrate), becoming a unique symbolic act without the need to communicate it explicitly or even to attach a definition to it.

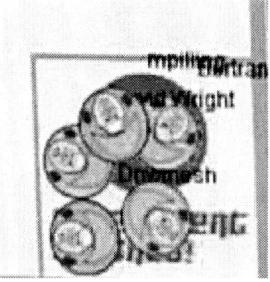

Figure 5.7 A spontaneous 'group hug' performed by four participants clustering around the facilitator.

These two cases are also examples of *design for emergence*: people explored the game movement and physics, discovering that by moving really slowly they can stick to another car rather than get bumped off. So participants came up with the 'group hug' through experimentation. In the second case, the two participants explored spinning around together as a means of expression.

We also observed some interesting forms of 'crowd' behaviour online. In several experiments, a most active individual emerged, much like a playground 'wise guy' who tended to ignore the game's context, for example, by driving around other people and trying to 'bump' them out of their parking places in the colour jam session experiment, which was the non-bumping variant of the game. Frequently, others followed this behaviour, too. Emergent play occurred: participants ended up spontaneously swapping places, sometimes even 'offering' their place to others! Several people mentioned the following as fun to do: whenever someone headed for a place nearby, they would come out in front of them to 'steal' it. Here is how one participant described in the questionnaire this behaviour:

> *Small games happening spontaneously, with no communication! – people start chasing each other. People try and get to a spot first, just because someone else is heading for it. (I did this twice)...*
>
> *Possibly, people mess about because it's fun to disrupt a group's activity – especially when there is no leader / hierarchy enforcing control, and there is no penalty for messing around, either enforced by the game or dealt out by players.*

At one point, in the second colour jam session experiment, a 'rogue' participant 'stole' the onscreen facilitator's place in the game and seemed reluctant to leave for several seconds (figure 5.8).

These emergent playful behaviours were amusing to observe, convincing us that there does exist such a thing as an 'online crowd', even within a simple 2-D bumper car game. It is important to point out that the same 'rogue' individuals were also creative and good innovators, taking up initiative in various contexts. For example, the same individual who 'stole' other players' places, tried later to introduce interesting complex patterns for others to follow in the colour jam activity. He started moving around his parking space in circles and demonstrated this pattern in slower motion as well. Several people followed his pattern, which he kept introducing frequently during the session.

Figure 5.8 A participant has taken the session facilitator's parking space.

Figure 5.9 The visual instructions in the second colour jam session, signaled by the facilitator's bumper car moving onto the appropriate icon. The second instruction is place swapping, incorporated as part of the session to liven up moments without much activity and to create variable teams, in a way that resembles a children's game of 'musical chairs'.

Our model of design for emergence for these experiments includes personal creativity and group dynamics (people imitating each other's behaviour) as external factors. Together with a simple design allowing for experimentation these factors led to the emergent behaviours of 'group hug', 'victory dance' and the aforementioned spontaneous place swapping. After we observed the place swapping people did in the first colour jam experiment, we decided to incorporate that activity in the game design itself in the second session with the two teams. Coordinated place swapping, indicated by the facilitator going to the appropriate icon (figure 5.9), worked well because participants got to try making rhythmic colour changes and coordinating with different individuals in various group combinations. Also, when their activity would slow down, giving an indication that people might be getting bored, it was a good way to bring liveliness in the environment: they would start bumping each other and being playful during those 'place swapping' breaks. This is one example (figure 5.10) of how design for emergence can work as an iterative design process: provide a simple design and observe the unexpected uses, identify engaging and interesting user behaviours and try to incorporate some of these aspects in a future design (in our case a playful online activity).

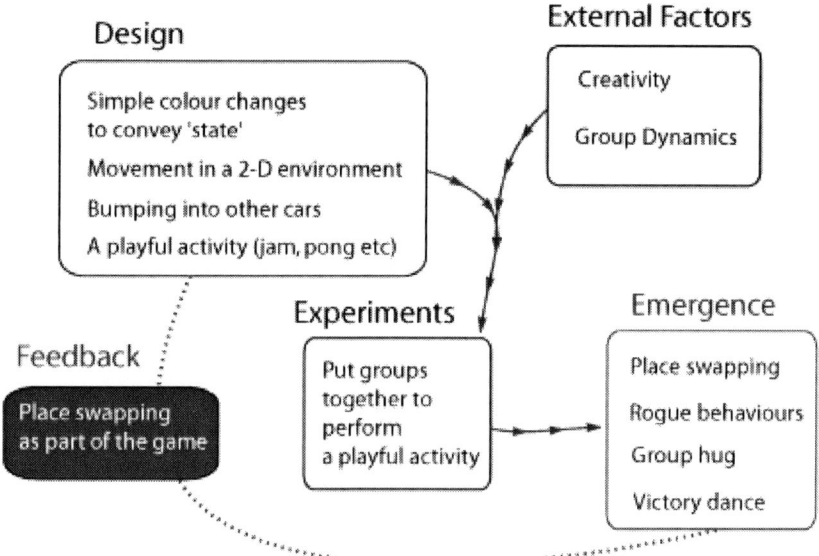

Figure 5.10 The design for emergence model with the BumperCar game.

5.5.2 Spontaneous collaboration and group behaviours

We observed spontaneous group behaviours and collaborative actions within the context of the three activities. In the colour jam session experiments, synchronised colour-changes and creative 'dance-like' movements were observed regularly among two groups (figure 5.13). Remote participants managed to engage in shared tactics with their team by observing what other people were doing. Visual clues based on colour and simple local rules that emerged on the fly (e.g. copy the colour of your neighbour) led to a more complex behaviour (everyone having the same colour or performing the same pattern) triggered by one individual, elaborated by another and ultimately adopted by all team members. While analysing the videos, we noticed an interesting progression: starting from a phase where people were trying to find out what to do by copying each other's colour, they finally came up with quick, coordinated colour changes and wave-like patterns with increased complexity in movement. In the first colour jam experiment for example, the circle shape encouraged people to ripple a colour around in a wave like manner. After the first six minutes without any signs of synchronisation or shared pattern, half of the circle turned the same colour, blue (figure 5.11). This was the first coordinated attempt, which happened because a majority colour emerged (three people were blue in a row while all the rest had random colours) and people followed it, one after the other, exactly with the order they were on the circle, clockwise. This indicates that the circle shape encouraged people to spread a colour around and that it is possible for such a coordinated behaviour to emerge without having to talk about it. That very first experiment was not successful because participants were too distracted using the chat and there was a lot of confusion about what exactly they were supposed to be doing, so not everyone followed on time. In the second colour jam experiment, half of the people (four out of eight participants in total) hardly used the chat at all as it was just a background 'emergency' communication channel. Yet, they would all pick up behaviour patterns, just by observing what other participants were doing. With the visual instructions added, that experiment was more successful than the first one and participants came up with many different coordinated movements.

While this is a very simple case, it indicates that real-time self-organisation can indeed emerge in online multi-user environments. Examples of similar self-organisation within the game context were noted in the other experiments as well. In group formations and chasing (figures 5.14-5.15) we observed swarm-like movement and alliances on the fly. People demonstrated different tactics, like following the larger team (occasionally this resulted in everyone being divided among two dominant teams/colours) or being a 'rebel' and challenging others to chase them.

One problem we observed in the two team formations experiments when correlating the videos to participant feedback, is that it was quite difficult to remain a member of the same team for more than several seconds. Players were sometimes chasing someone and then suddenly would change their colour to become team mates or vice versa or abandon people that tried to keep up with them. There was a sense of instability, because teams were temporal and unsustainable. The reason for that was mainly the physics model: 1) the growing distance between cars when driving 2) the absolute freedom to move towards all directions, and 3) the bumping effect occurring every time a team member bumped into a member of the same team accidentally.

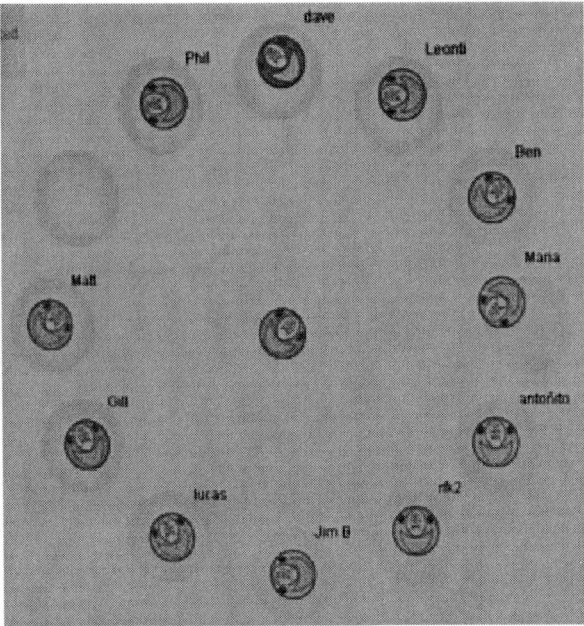

Figure 5.11 Players spreading the blue colour around the circle clockwise.

We therefore have come to the conclusion that groups of players should have a different physics model than individual players to allow for more sustainable groups to be formed. This would involve a significant redesign of the prototype with a clear focus on groups. We believe, following earlier discussion, that enforced social hierarchy and mechanisms for leader nomination can be a problem, because of their complexity. A more open, 'democratic' solution would be to introduce a model that calculates the average movement of a group from all individual movements, more like a swarm in which everybody has some freedom of movement but not so much as to lose the group. We would also suggest semi – permanent teams, to allow some fluidity and flexibility and keep participant's interest in forming groups with different people. For example, if you join a team you would need to stay with it until a given time when teams are allowed to exchange members again, like in the second colour jam experiment where we had semi-permanent teams with team swapping. This would possibly allow for group strategies to emerge.

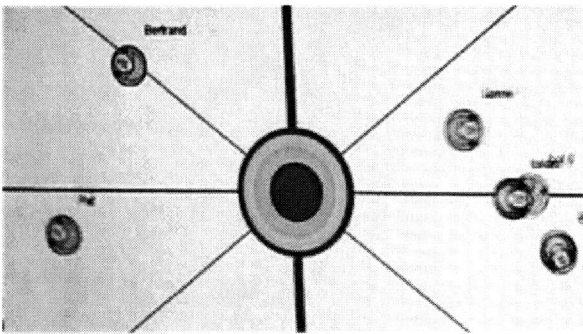

Figure 5.12 Upper/lower space divisions among members of the green team on the left. Also a role division in the same team, two members are 'defending' and one member has entered the other team's territory.

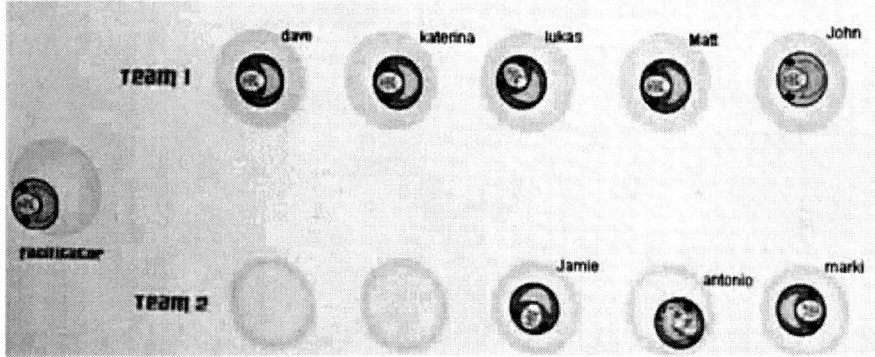

Figure 5.13 Same rotation angle and gradual colour changing

In the first collaborative Pong game session, participants spontaneously turned it into football, which has quite a different style of play. Instead of staying on either side of the screen to bounce off the ball, they tended to follow the ball and cluster around it all across the game area. All seven participants found the game engaging—they enjoyed this 'school boy football' style, 'being in a team' and 'the strategic aspect of the game'. One participant reported in the questionnaire that he liked the fact that his team seemed to split automatically into 'defenders' and 'attackers' without having to talk about it. This is an example of emergent, unplanned collaboration – based on the awareness of other people's activity. Other, frequently observed, forms of ad hoc collaboration in this game were space divisions: team members often tried to keep a distance from one another dividing their part of the screen into front/back or upper/lower to make sure they covered as much territory as possible. See for example in figure 5.12, the Green team (the two cars on the left) has an upper/lower space division while their third member is on the attack. Again, the only way people could come up with this was by observing what others were doing: there were no such explicit instructions, strategies or communication within the game.

The experiments demonstrate that it is possible for ad hoc, spontaneous group collaboration practices to emerge spontaneously with purely visual communication, just by observing or being aware of other people's activity. Participants received instructions about each experiment, which helped provide a necessary context about the activity (jam session, group chase and Pong) but these also allowed freedom of expression and space for group coordination on the fly.

The types of emergent behaviours we observed, both within the context of the playful activity and completely unpredictable ones, are summarised in the table 5.2:

Table 5.2 Summary of observed behaviours

Behaviour	Examples
Subvert the game, be a rogue (often resulted in 'crowd' behaviours)	- Move around bumping others, mess about - Place swapping, offer or 'steal' a place - Steal facilitators place
Try to be a leader-innovator within the game context	- Initiate wave like colour changes or insist on one colour - Move around your space in a pattern
Improvisational, expressive group performance (spontaneous, emergent teamwork)	- Spontaneous Group Hug - 'Victory dance' pattern - Football style 'Pong'
Goal-based teamwork	- Surround a target with others - Divide space to upper/lower or front/back - Take up roles spontaneously: defenders and attackers - Form a block with team members to bounce off the ball
Self-organization	- Keep up a wave of colour change and movement until the neighbours copy it - Copy the colour changes and movements of neighbours - Join other cars of the same colour on the fly in a chase
Trying to draw others' attention	- Be chased or try to be chased - Change colour to make others follow you - The 'rogue' behaviours above

5.5.3   Game Experience

We identified the following factors that influenced the players' experience:

1) The social factor: 'misbehaving' and being a 'rogue'

The social aspect of those activities was influential: people enjoyed creating their own fun with others. What participants found most engaging in these game experiments was their playful interaction. This became clear from the first two experiments with the colour jam sessions, where sometimes participant behaviour resembled that of children in a schoolyard! In the first experiment four out of seven people clearly stated that they enjoyed it when themselves or others were moving, messing around. They also enjoyed 'rogue' and playful group behaviours (e.g. the spontaneous 'group hug' – when they all clustered on one spot). In the group formations experiments, a couple of people mentioned they liked chasing individuals they knew or they 'put on the spot' and another two said they really enjoyed being chased by others, especially by a group of people. In the most goal-oriented type of game, the collaborative pong, engagement was associated with pursuing a goal with others and collaborating as a team. In the first Pong experiment, all participants (seven out of seven) said they enjoyed the game as a whole.

From both videos and questionnaires it became evident that 'messing about' emerged when people felt a bit bored, when they did not know exactly what to do and when there was not much action on behalf of the others. For example, in the first jam session we observed that as soon as attempts for colour patterns and coordination failed or when there was not much activity, somebody would start bumping others or try to get their places etc. In the second Pong experiment, a participant who mentioned he liked *'subverting the game by bashing into other players and breaking the rules'* was the one who enjoyed that particular game session least. This reminds us of real life crowd events and the Mexican Wave in particular which emerges when the football game is not very interesting to watch or during intervals and periods with not much activity (Farkas et al, 2002). Similarly, the unpredictable, spontaneous meta-games we observed online emerged when people were bored or less active.

2) Clarity of the perceived activity, purpose or context

We found that lack of clarity in respect to a perceived goal or purpose within an online playful activity also affects motivation. Our very first colour jam experiment was the one least enjoyed as several participants mentioned they felt frustrated because they were uncertain and confused about what exactly they were supposed to do. This feeling of 'being lost' affected negatively their motivation. They were being distracted by the chat channel and switching attention between the two made synchronisation very difficult. In the second experiment, patterns emerged more easily as they were focused on the environment and people gradually became more active because they had a clearer perception of what was happening and what they were trying to do. This session was more coordinated than the first experiment and people produced complex patterns collaboratively. They also enjoyed it much more, as we can suggest from comments made in the chat room after the end of the session (e.g. *'That was a good laugh!', 'n every body go CRAZY!!!!!', 'lol'[which stands for 'laugh out loud' in chatroom language] 'much fun :)', 'aven't you lot got homes to go to?'*).

In online environments, where social cues, body language and expressions are limited, people have a greater need for feedback in response to their actions. As participants were unsure about the 'criteria' to judge their performance in the first colour change jam session, they got confused and their motivation decreased. The improved layout with instructions signposting different phases and relaying less on the chatroom helped them get a clearer idea of the event. In the other experiments, in collaborative Pong and group formations, participants wanted to have performance feedback, in other words to know at any point how well one is doing in the game and which team is better. In this context, the aforementioned 'victory dance' that the two participants came up with, was their own invented feedback to themselves, indicating that they are coordinating really well as a team. Because teams in group formations were constantly changing, they also needed immediate feedback as to who was part of their team and who was not and where they were heading towards. While typical game performance feedback (scores etc) is not a priority in the kind of free-play we are investigating here, we still need to take feedback issues into account when designing engaging social experiences.

3) The number of participants and group dynamics

Our experiments showed that even a small increase in the number of participants in these small groups can make a difference in the emerging interaction and result in a variable user experience. For instance, with more participants in group formations and chasing, video analysis revealed that larger teams (three or more), chasing a car of a different colour, were more successful than pairs. These teams managed frequently to trap or 'squeeze' a target, as there were more people to cover gaps and to catch up with a chase. In the second group formations experiment (which had 8 participants whereas the first one had 5), larger teams were also most successful (teams of 3, 4, 5 even 6 persons). What we noticed in the first experiment was that when a team of 3 members would be formed, one out of three cars would be usually left behind quickly and there were not enough people to cover 'gaps' or just join the team ad hoc. This was not the case with eight participants, where we observed teams of 3 or more being formed spontaneously. Three out of four people in the first experiment confirmed in their feedback that groups of 2 worked better for them. In the second experiment, four people out of eight said that teams of 3 worked best. There was only one person who participated in both and could therefore compare the experience between the two. After the first experiment with five players, he said he preferred being in a small team of two. He changed his mind after the second experiment with eight players and concluded that larger teams work better. He also mentioned about the second experiment: '*I liked it more than the previous, I was almost all the time in a team of four and sometimes we managed to chase cars together*!' This indicates that even a small increase in the number of participants can make a difference in the interactions that will emerge.

Another example is the difference between the two collaborative Pong experiments. The first experiment with seven participants turned out to resemble more a game of football (figure 5.16), as people moved altogether around the ball like children do in a playground. However, the second experiment with four participants proved to be more like the actual game of Pong, because participants were more defensive, staying most of the time on their side' of the screen waiting for an opportunity to bump (hit) the ball. One participant said the first experiment had a 'schoolboy football style' and another said it reminded him of 'ice hockey'. In this experiment a participant started moving around very freely in both teams' sides from the very beginning. This established immediately a free, spontaneous use of space without anyone having to set any rules about where to go and what position to take. The second Pong experiment did not have this atmosphere, as players stayed mostly at their side. Even after the first 7 minutes of the second experiment, when a participant started performing a more active, offensive strategy by invading the opposite team's territory and attempting to send the ball to the opposite team's border, he would still go back to his team's side as soon as the attacks either failed or succeeded, because there was a lot of space to cover. Also, given that the ball moved a little faster than the players – it was more efficient, for a small number of people, to wait for an opportunity, rather than chase the ball around altogether. In contrast, in the first experiment it was possible to move around in a swarm-like manner since there were more people to cover different directions.

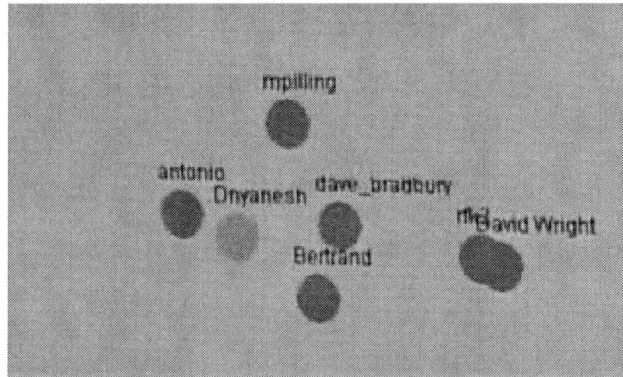

Figure 5.14 A large group has been formed and players try to surround a lone player.

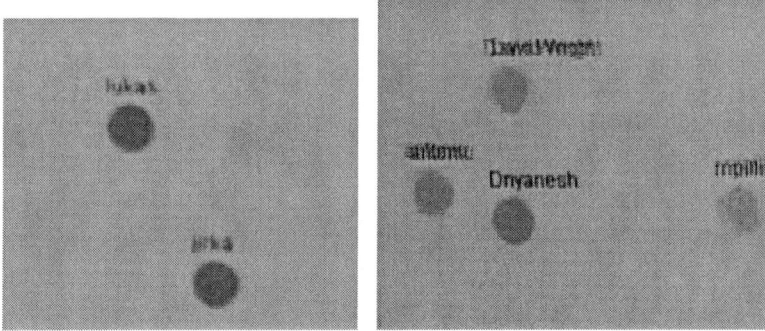

Figure 5.15 A pair in the first experiment and a group of three in the second experiment

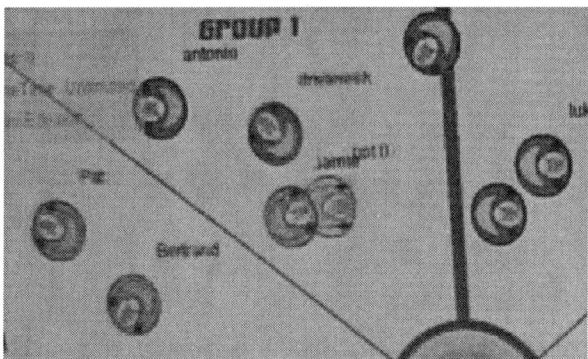

Figure 5.16 A game of collaborative pong turned into 'schoolboy football'.

Participant number is not the only factor that influences the player experience and emergent interactions, but it is an important one. People's perceptions, expectations and game experiences also influence their behaviour and often subsequently the group dynamics of the game. So, for example, one participant in the second, pure Pong experiment mentioned that the opposite team should not have entered her territory, even though there were no such rules or restrictions. She never attempted to enter theirs and she was not happy with the other team's invasion.

Another example where the size of a team seemed to make a difference was in the second, team-based experiment of the colour jam sessions, where people tended to form uneven teams many times (e.g. one full team of five and three people left out at the opposite side), despite the fact that they were instructed to distribute themselves evenly among two teams. Also a couple of times one participant abandoned his team and moved to the opposite one, even if there was no more space and he would just stay between two cars (figure 5.17). There was clearly a tendency to go with the majority whenever possible, so we interpreted it as being more rewarding to be part of a larger team, with an advantage over the opposite small team. Also, competent players attracted more people around them, forming a large team quickly. A couple of participants commented in the questionnaire: *'the more people in the team the better the pattern looked'* and *'at first it was random and people were unsure where they were going, but when people figured which players knew what they were doing they followed them'*.

Given our commitment to designing scalable multi-user environments, our conclusion here is that there is a critical mass for certain interactions and behaviours to emerge and that the number of people present affects the user experience. For instance, people in the football-like collaborative Pong session enjoyed it more than people in the rather pure Pong session did. This was a study involving small groups of people; in future work, there might be more emergent social phenomena yet to be discovered on a larger scale.

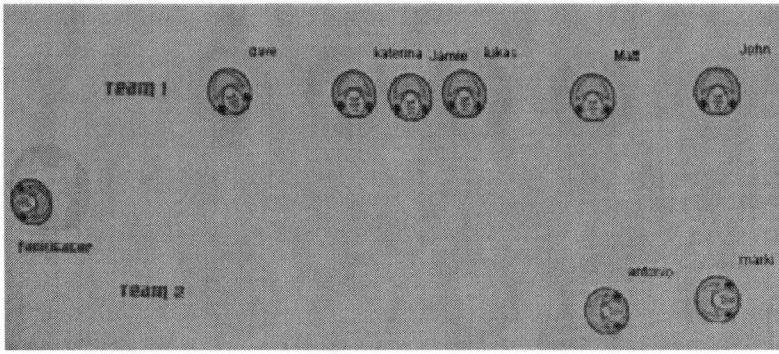

Figure 5.17 Two uneven teams have been formed and a participant from the small team has moved between two players of the top, larger team, even though there is no 'parking space' left, indicating he wants to be with them

5.5.4   Visual communication and design

We focused on visual communication within the context of these playful activities because it is more immediate and scalable than textual and verbal forms of communication. If we want to create playful participatory activities to engage large numbers of people in the future we need to find the best and simplest way for them to communicate. Here are our observations from our six experiments on visual communication, design improvements and the use of the overview map.

1) Visual cues proved very useful for providing context on the activity

As already mentioned above, in our very first experiment, we ran a separate chat facility alongside the game, in order to coordinate the start and end of the experiment, and also to allow people to discuss strategies and make suggestions about their performance in the game during a specified chat break. We found that it was particularly difficult for people to keep up with the game's activity and at the same time pay attention to instructions or conversations in the chat room. The chat was too distracting, particularly when people were trying to come up with something altogether, in our case a rhythmic colour combination. Quite often people were not paying attention at important moments, causing disappointment to the rest who were synchronizing. As one would expect, a chatroom is not scalable and it is really easy to miss an important line from the experiment facilitator or another player when 10 people use it altogether. Also, in this first experiment, five people reported in the questionnaire that they felt 'lost' in the beginning of the session because they or others were not sure about what they were meant to be doing. For this reason we improved coordination by providing the aforementioned visual cues about each 'stage' of the session (figure 5.9). The facilitator used the visual instructions to tell people when to 'stop', 'start', 'swap places' etc rather than chat lines. We observed that these were successful as participants responded and followed the context of the session without problems and were much more active and involved. Most of the people commented the visual instructions worked well and one participant said he would like to have a small text message popping up in the game in addition. Thus, visual communication in such an environment is more crucial and efficient than verbal. In the second colour jam experiment the chat was retained as an 'emergency' communication channel in addition to the visual instructions. Two or three people used it to make jokes and suggested a couple of ideas about interesting moves. The rest five-six participants did not focus on the chat at all, but just observed what others were doing and followed their moves and patterns accordingly. In fact, five out of six respondents said that observing others was more helpful than the chat in their attempts to coordinate with team members. The chat was eliminated from the following experiments and only visual instructions were used, even if they only represented the 'start' and 'end' of an experiment.

These observations make a point on the organisation of such online events with distributed participants: clear visual contextual cues are more appropriate than chat to signpost each phase or specific activity during the online event (e.g. wait for others, start, stop, have a chat break etc).

2) Signaling based on colour flashes is very easy to miss and not efficient for group coordination in online environments

We observed in all colour jam and group formations experiments that colour changes and movement must be clear and lasting long enough to be seen. For example, in the beginning of the first colour jam experiment again, people kept experimenting with complicated colour changes. They did not try to do something simple at first and for long enough to be noticeable, so that others could follow. Instead they tried sequences combining all four colours together, which were difficult to follow. It was also almost impossible for other participants to keep up with each other as they hardly ever repeated the combinations they did. When their attempts were not leading anywhere they would stop in hesitation, some focusing to the chat. At this stage, participants had focused on their own performance and it was only when half of the circle turned blue that they started trying to coordinate. Particularly colour flashes were difficult to follow (e.g. from orange change to blue then back to orange) because they could easily go unnoticed by the next person who needs to keep up the pattern. Responses and durations of colour flashes varied for each person and this created confusion as some people were too quick others too slow, or there was a bit of lag and some were not switching back to the original colour. Several people flashed their colour before their turn, thus generating more confusion about 'who is next' and 'what are we trying to do'.

Similar problems were observed with group formations, where we suggested that players use a colour flash 'signal' to indicate to others that they want to team up and to draw their attention. We found that this idea for a 'signal' did not work in both experiments, neither as a prompt to form a team nor as an invitation to follow along in a chase. The reasons were that participants were not getting response for their signals (i.e. another signal) or they would get contradictory and confusing behaviours as a response, for example being followed but then abandoned or not being followed at all. Sometimes other people would respond to their invitation to team up than the ones intended, as it was hard to tell sometimes who the 'signal' was directed to. We found that just by changing the colour identity was enough to form ad hoc groups and chase together, so people tended to follow people of their colour or just change their colour as an indication that they are joining someone of the same colour without the need for a special signal. The only case we observed where the colour flash worked as a signal was in the first experiment, among the two participants who also did the 'victory dance'. One of them signalled a few times, possibly as a confirmation for still being a team and the other would respond with another signal and they would then keep moving together.

So, drawing on the above observations we believe that we need to have the following as rules of thumb for our environment, to ensure successful purely visual communication among users:

- a) Clear and simple state changes (in this case colour, but could be icons, shapes or other visual elements) that
- b) Have a minimum duration (e.g. several seconds). Colour flashes that happen quickly are very likely to go unnoticed.
- c) If a state change is complex (e.g. combination of two or more colours or a pattern of moving around one's space in circles), it must be repeated until the other user responds.
- d) When performing simple state changes (e.g. change my colour once), one needs to always wait for a response from another person (e.g. for him/her to change his/her colour) before changing their 'state' (colour) again to something different.
- e) A response is necessary to avoid communication breakdowns and misunderstandings. The response should also be as immediate as possible, following the original prompt without much delay.

3) It is possible to make assumptions about participants' behaviour expressed visually

The aforementioned example of the individual who started a complex pattern, which was then copied by others, illustrates the importance of group dynamics and individual contributions in every situation. With different people we might get different behaviours and interactions. Skills (driving in games, gaming experience, chat) and personality attributes (e.g. self-assumed leaders) are very relevant and can influence the experience the group will have. Although communication channels are restricted in the game, events are 'open' to interpretation and it is even possible to attribute certain qualities or personal characteristics to people by observing their behaviour, expressed through colour and movement in relation to others in the context of such games. We can observe individual performances and make assumptions about others' innovation, creativity, ability to collaborate and leadership, being team players etc. In our example, our participant seemed to undertake a leadership role: he tried to take the lead by showing his driving skills and initiating patterns. Occasionally, when people did not notice or follow his patterns he would start bumping them as if punishing them! Alternatively if he received no response at all, he would sometimes join the opposite team. Other people tried to attract others' attention in different ways: for example, a participant in the second group formations experiment, changed to a different colour every time there were two dominant colours to see whether others would follow. Another participant mentioned in follow-up feedback that what he really enjoyed and found fun, was being chased by many others and that he tried to keep a different colour and challenge other players to chase him by driving really close to them slowly. It was also possible to spot individuals collaborating, allowing others to lead, but providing effective support. Another case was that of people who worked well together as a team. For example, two participants in the first group formations experiment performed many coordinated chases together, exchanged colour flash confirmation signals, waiting for one another, and also came up with the aforementioned creative 'victory dance' pattern. Not to forget that the creative emergent behaviours mentioned above, like being a 'rogue' and performing a 'group hug', also suggest personality attributes. These are good examples of using the game as a means of personal expression.

After the first colour jam experiment, a participant identified different roles among other participants he encountered for the first time in the game: listeners, leaders, organizers and 'messabouters', providing interesting comments for each category. Other people also made comments about other participants who they did not know personally and discussed their behaviour in the follow-up feedback.

Therefore our experiments revealed an interesting dimension: if it would possible to identify some aspect of people's personality and social skills through simple games, then these could be used in different contexts, for example, as a team building exercise. To summarise, we believe that the following are minimum requirements for someone to be able to assume personality elements of other people by observing their behaviour in a simple, presence-based environment, especially if the number of participants increases:

    a) The ability to notice, to identify individuals. To be able to differentiate their behaviour in relation to other users of the environment and not get 'lost in the crowd'. Therefore, each person must have a presence.

b) The possibility to use design elements for personal expression. In the case of the Bumper Car game it was colour and movement that people used in expressive and creative ways.

c) A context of interaction. Some kind of collaborative task, game, competition, performance that will allow comparison between individual contributions and behaviours.

4) 'Overview' versus 'close-up' view: a personal choice

As mentioned in paragraph 5.4.2, the map view was used only in the two group formations experiments. In the rest of the experiments the close-up view was sufficient since we only had small groups of people and a fixed space for each activity, so there was no need to use a large overview map. The map is a significant part of the original game design, aiming to promote scalability and a sense of 'crowd presence'. The original idea was that with an appropriate map view we could increase potentially the number of participants to explore to what extent we can involve as many people as possible, while keeping the balances of visibility and playability. Most people in both group formations experiments used the map view, but stayed most of the time in normal view. As we had small groups and subsequently small environments, only a relatively small part of the whole view was 'out of range' in the close-up view. So, one could spend time mostly in the normal view, occasionally switching to the map to locate others. Each player's self-view indicated driving direction with a small arrow, but steering was a bit more difficult in the map view because of the small angle. Two people from the group formations experiments (out of twelve in total) commented in the questionnaire that it was not very convenient to use the map all the time. It was more difficult for them to steer and to 'determine direction of others', so they preferred switching between views. One participant, however, said that he stayed in the map view all the time as he found it easier to move around and see where all the people are.

As minimum requirements of the overview map, a) it should be obvious who the user is within the environment at any time and b) the direction of movement should be visible, ideally for both the user and other players. A future, improved design of the map would include a direction indication (line or arrow) for everyone in the environment, not just for the user. We would then need another, more explicit way to identify the user's car quickly (e.g. a circle). There is a trade off between advanced functionality and complexity, so the design would depend on the context of the game. We are in favour of switching between two views rather than having a typical small radar screen at the corner of the interface as in many MMRPGs (e.g. Asheron's Call). The reason is that the full map view can create a more immersive sense of a collective presence, knowing 'who else is there with me' and moreover, it can be more scalable as the number of participants increases.

5) We need to provide more visual 'presence' cues

Further to our design principle about presence being 'symbolic' we found that knowing others' attention/ idleness is vital for collaboration, particularly in non-verbal environments. There were several illustrative examples in our experiments. As we already discussed, in the very first experiment, the chat was very distracting when people were trying to come up altogether with a rhythmic colour combination. Quite often participants were not paying attention at important moments, causing disappointment to the rest who were getting synchronised. In a hybrid environment involving both text based communication (a chatroom) and a visual 'playground' (the actual game) we would need an indicator of 'where' a user really is, i.e. chatting or playing, in order to communicate the users' state of attention.

In another case, a participant reported in a follow-up discussion that he stopped playing the collaborative Pong game for several seconds when someone came to his desk to talk to him. People in the 'playground space' could not notice the player's lack of participation immediately. Video analysis revealed that a collaborative attempt failed because that person was 'idle'.

But presence is not just about the state of someone's attention. In both group formations experiments we had cases of bad team coordination, because of delays in player response, but most important because there was no way to tell who is *really* in your team and how long they will be there. Even though there were cars with the same colour very close, they could still move in a different direction or stay back as there was no way to know their *intention*, only by observing their moves. It was also difficult to identify the target every time. If there was one car of a different colour among cars of the same colour, the target was self-evident, but this was not always the case. We suggest some design enhancements that would improve on this front in our conclusions next.

5.6 Conclusions

Our experiments with the BumperCar 'playground space' illustrated how collective spontaneous behaviours can emerge through online playful activities, based on symbolic presence, visual communication and affordances. Some of these behaviours were unexpected, using game specific features for emergent social play beyond the introduced game context, a kind of meta-game (e.g. 'group hug', 'dancing' in a pattern). Other group behaviours we observed involved ad hoc collaboration within the context of the specified activity (e.g. swarms, seeking protection in larger teams, wave-like colour patterns). Our participants enjoyed these spontaneous group interactions and their self-organising performances, whether within or beyond the game context. We have come to the conclusion that we can only *facilitate* spontaneous group play rather than *evoke* it in this online environment.

The presence of some kind of authority or rules which players can challenge or subvert has been identified as a significant factor in steering unplanned individual and group playful activities. In our experiments there was a facilitator coordinating the start and end of the session and certain other phases. While not exactly an authority, the facilitator can be seen as a 'schoolteacher' in the playground space, a person who can be challenged. The existence of simple rules that people can break or subvert (e.g. no bumping others, no talking) often evoke interesting behaviours. Breaking rules (e.g. all the 'messing about' we recorded) can be fun, a sort of 'illegal' entertainment, as long it doesn't continue for too long and disrupt other people's activity. In other words, the system or game could *allow* the user to break the rules in some way. For example, if the rule is to stay for some time in a parking space, a player can break this rule by moving about. This is how a whole meta-game started with those parking spaces, reminding us of 'musical chairs'. In other contexts though, it can be hard to draw the line between allowing people to have fun by breaking the rules and completely disrupting a group activity by being a 'rogue'.

If breaking the rules is not particularly desirable, the environment should at least provide some tools for people to use creatively. So, another alternative is to provide for *freedom of expression: the ability to experiment, do other things beyond an assigned activity, which are integrated in the design.* This is perhaps the most important ingredient. The Bumper Car concept, based on a familiar metaphor from a real world experience to which most people can relate, provided *affordances* for an exploratory approach, encouraging users to try and experiment with the physics of bumping and moving around. In this case, colour, movement and the ability to collide with other players immediately generated a fun element. Imagine if, instead of static lists of names and dots in an Instant Messaging application, you could move around and bounce, like in Bumper Cars, what would happen?

Our experiments confirmed the importance of symbolic presence. The breakdowns in coordination we observed, illustrated how vital it is to know other people's attention and mental state ('Where am I really?'). In our games, an automated 'attention state' would be valuable, for example when a user doesn't press any keys for several seconds, their car could become 'grey' or fade gradually. At the same time we suggest that the 'attention' state should also be user-defined, so that the user can turn to 'idle' mode with a quick key press. In this way, if I am just going to pick up the phone or get a sandwich, I can quickly change my state to let others know I am unavailable. Furthermore, there are various other aspects of a person's 'mental state' that can be crucial in the context of such games and environments. We have identified team membership, intention and activity (e.g. current direction of movement) as important states for collaboration and coordination within the group formations experiments, an attempt to encourage flock-like behaviours among groups of players. Our existing design uses colour to convey team membership, but further enhancements are required to facilitate team coordination as well as to maintain consistent teams. So, for example, in group formations we should provide users the facility to highlight their target for others in their group to see (intention).Visual aids (e.g. arrows, paths) that highlight a person's direction of movement can be helpful to promote team play, possibly also restricting the choice of directions to four so that people can follow along easily. We know that the concept of symbolic presence acquires meaning from its particular context, thus different presence attributes are required for different games and social environments (e.g. consider Instant Messaging versus Bumper Cars). Following our principle of *lightweight design* in chapter 3 and our aim to provide just enough experimental space for people to explore and interact spontaneously, we feel that, as a general principle, presence information should be symbolic and as simple as possible, but sufficient to support awareness of other's activity.

In addition to our four design principles, we believe that when designing a multi-user online environment we need to provide *a context and appropriate feedback for group interactions*. We found that it is not sufficient to put a group of people together; there needs to be a specific reason for being there or an activity to which everyone can relate. We can break down the concept of collaborative, social play to fundamental ingredients, aiming to foster group interactions online:

a) A context. This can be a specific goal, as in our Pong game, where we had two teams and each was defending one side of the screen. However, unlike most computer games, the context of interaction need not necessarily be defined by a goal; it can be open-ended, much like unstructured play in the school playground. Whether bumper cars, Pong or rhythmic colour change competition, it is important that the context provides *affordances*: people are able to associate the online collaborative activity with a real life experience, for example with a music jam session.

b) The ability to perform certain acts within the context. People expect to participate in some way in an online environment; in our example, using movement, colour change, following others, chasing others, bouncing off etc. In real life when people gather to socialize for example, they talk to others, make gestures and use body language.

c) Feedback: both individual and group. Our participants emphasized that they needed more feedback from the game and from other participants in response to what they were doing. This feedback was necessary both at the level of *confirmation*: 'am I doing the right thing? Is that player still with me? What are we doing now' as well as for *performance*: 'who or which team is better?'. The first level of *confirmation* can be addressed by introducing contextual presence states, communicating team membership, intention and activity, as mentioned above. A reward mechanism, for both individual's and team's performance, can satisfy the need for *performance feedback*.

Through our experiments we also found that it is possible to *make assumptions about aspects of people's personality and social skills* through simple presence based games, in a minimal 2-D bumper car environment, with no verbal communication, only colours and movement. As we saw in our experiments, there were several occasions when people used their presence in the game in a creative way, as a means of expression and spontaneous social play and meta-games emerged. In these situations, it was possible to attribute social skills and personal attributes to people, even complete strangers. We found that activities requiring collaboration and observing other players' actions in the game often give away hints about their social skills, such as whether they assume the role of the leader, innovator, team player, supporter, lurker etc. In this respect, it is also important to be aware of individual innovation. All emergent group behaviours we discussed earlier started with the innovation of one or more individuals, followed by others. It is important that people can draw attention to themselves and make others aware of their activity, especially if user participation scales up.

In our model of design for emergence above, we identified *individual innovation and creativity* as well as *group dynamics* as external factors that influence the emergent interaction. Complex patterns as well as 'rogue' behaviours had to be performed by someone before spreading to the whole group. We would need to try these games with different kinds of audiences and age groups in order to gain a broader range of emergent behaviours based on the different group dynamics. We observed various interesting emergent behaviours (place swapping, rogue, group hug, victory dance) in the experiments, showing that it is indeed possible to design for emergence and to be surprised by people's unexpected uses of the design. We also integrated one of the emergent interactions, place swapping, as a game activity to facilitate the formation of variable teams and to add fun to the less lively moments of the session. This was the feedback loop of our 'design for emergence' model: emergence inspiring design. There is a limitation in this approach though; not all emergence can be translated into design and the idea of including an additional feature for something that people already do spontaneously might spoil all the excitement and fascination about it. More interesting would be to explore in future work how our lessons learned from the process of designing for emergence can be applied in other contexts and inspire the design of non-game focused social interactive systems. We further discuss the feedback of emergence in the design process in chapter 8 of the thesis.

One of the greatest challenges for the designers of collaborative and community support systems is to cater for the vitality and spontaneity of human interaction. We believe that play, as a fundamental aspect of human nature, can illuminate some of these issues and help us learn from the 'playground'. We hope that the results from our studies, and in particular our design elements for emergent play and collaboration, will inspire designers of collaborative systems and multi-user environments. Such playful experiences along the lines of our organised online activities appear to be beneficial for community development, with the aim of creating and enhancing bonds among distributed individuals. They could potentially help online participants to identify good team players and establish a point of reference for people to meet and get to know each other. There is an opportunity here to leverage social skills that could be applied in the design of future applications for collaborative work, learning, play and social software by providing means of personal expression and group activities. One direction for future research that could follow on the principles established in this thesis, is to apply playground experiments of this kind in non-game related contexts, for example for community building in distance learning environments or team building among small groups. We hope that future research work can extend these experiments with enhanced game prototypes to explore how emergent participatory play can spread the feel-good factor of 'being together' among distributed individuals in large online communities.

In the next part of the thesis, Part III, we explore collaborative social play as an experience expressed in the real, physical world, mediated by a superimposed virtual situation. This work takes into account the opportunities and implications posed by the development of emerging technologies and the concept of ubiquitous 'presence'. The research focuses on whether the virtual world can motivate or encourage the emergence of spontaneous group and individual behaviours in the real world. Based on lessons learned so far, our design process attempts to blend mediated collaborative play with the fabric of everyday life.

Part III

Design for Emergence in the Real World: Experimenting with a Mixed Reality Urban Playground

6. CitiTag: urban space as a large group playground

6.1 The idea

The work discussed here follows from the scenarios and storyboards on mixed reality games and playful activities we outlined in chapter 4 (e.g. PixelTag) and focuses on the notion of 'playground tag', an inspiring paradigm throughout the thesis.

The research objectives were:

- To understand how design can encourage emergent group behaviours and playful interaction in public spaces and how these social experiences can be motivating and compelling.

- To explore how people can use creatively their virtual and physical presence through a mixed reality game.

- To design an experience that could blend in with daily life activities.

- To incubate emergent values in the design with users being co-creators of the experience itself. We wanted to see how our concepts would evolve through their use.

The aim of the experience was to awaken the inner child within us, encouraging spontaneity and instant fun, based on the same principles that have guided our designs in the previous online studies, but in the context of ubiquitous, mixed reality interaction. From the extensive analysis of Opie and Opie (1969) we chose the following variations of playground chasing games as possible ideas for our mixed reality game:

1) *Pure Tag*. This is the very basic, original game of 'tag', where one person is 'it' and he or she tries to pass the 'it' to someone else by touching them (figure 6.1).

2) *Viral spread of 'it'*. In this case, the 'it' is like a virus that more than one chasers pass around by touching others (usually it starts from one person), until all people have been 'touched' and have become 'it'.

3) *Link hands tag*. In this variation of tag, the chasers have to hold hands when chasing others who have yet not been tagged. As the chain grows larger it becomes increasingly more difficult to chase (figure 6.2).

4) *Capture the flag type of games*. Massively multiplayer online games like Quake originate from this game. Here, the concept moves beyond the idea of simple tag the players being divided into two teams, with each team trying to steal a token from the other team. The game involves trying to work out strategies to invade the opposite team's zone and steal their 'flag'.

5) *Variants with chasers and runners*. Classic 'cops and thieves' type of games fall into this category. A very common chasing game found everywhere around the world is where one team is trying to catch the other and members of the 'runaway' team are trying to free their team mates who have been put into 'prison'.

Figure 6.1 Children in the playground making a draw (in Opie and Opie, 1969, picture taken by Fr Damian Webb OSB, © Ampleforth Abbey Trustees, permission granted).

From these variants of chasing games we found the last one most appealing, because by having two opposing teams the result of the game is unpredictable and players have to team up and collaborate, therefore play can become more interesting and provide space for emergent group behaviours in an urban context. A 'capture the flag' variant could be added on top of this basic chasing game at a later stage. However, in our game we decided to give exactly the same role to each team, therefore members of both teams could either be chasers or runners, depending on the situation they were in. We wanted to have more unpredictability for the final outcome and to introduce some social dilemmas in the game along the lines of 'should I try to rescue my team mates or go and chase my opponent?' in order to encourage emergent cooperation. We also wanted to accommodate the concept of linking hands from variant 3 above in our design, which appears promising for the kind of emergent cooperation we are interested in. So we tried to speculate how linking hands would translate in the virtual world and how it would actually encourage people to converge and cluster in groups in an urban environment.

In our design we tried to specify the minimum rules and structure possible for people to understand the game and its metaphor, and for cooperation to emerge. The city space then would become a playground and passers-by the usual or unusual suspects in a novel experience. We named our game CitiTag in reminiscence of the original game. Our initial storyboards and scenarios are briefly presented next.

6.2 Storyboards and scenarios

6.2.1 The action of tagging

One of the first problems we had to solve with the two way tagging, in which anyone could either be a 'chaser' or 'runner', was *how to decide who tags whom*. So considering an encounter of two people, each from an opposing team these are some ways to resolve the conflict:

1) Whoever presses a button (or other interaction input) first is the 'tagger'. This solution was not favored initially as it relies on computer skills and resembles location-based shoot-them-up games like Botfighters.

2) The tagging would depend on proximity boundaries. One would need to get close to someone to tag them, but not too close to invade their 'personal space'. For example, consider an area that has been divided into squares or circles. The person who is on an adjacent square (or circle) to someone else, would be close but not too close, so he or she would be the winner of a tag event. However, if he or she got too close and ended up on the same square with the opponent, the opponent would win. We liked this idea of exploring boundaries of closeness or proximity, but unfortunately in a real implementation it would get very messy and confusing with GPS inaccuracy.

Figure 6.2 A variation of tag with multiple chasers who need to hold hands when chasing. Only the first and last person can tag others (in Opie and Opie, 1969, picture taken by Fr Damian Webb OSB, © Ampleforth Abbey Trustees, permission granted).

Figure 6.3 Another variation of tag: the ball becomes the extension of the chaser (in Opie and Opie, 1969, picture taken by Fr Damian Webb OSB, © Ampleforth Abbey Trustees, permission granted).

3) Various battle ideas were considered as a means to decide upon the winner of a 'tag' encounter (e.g. strength, skills, educational quiz etc). The problem with this approach is that the focus gets shifted to warfare or competition and we lose the simplicity of tagging.

4) The last idea was based on the notion of specifiable locations: there would be areas friendly to one group (where people belonging to that group would be at an advantage and win a one-to-one battle) and areas friendly to the other group. People would need to discover these areas by moving around and trying to tag (figure 6.4). Colour differentiation or shadows on their display could indicate such friendly/unfriendly zones.

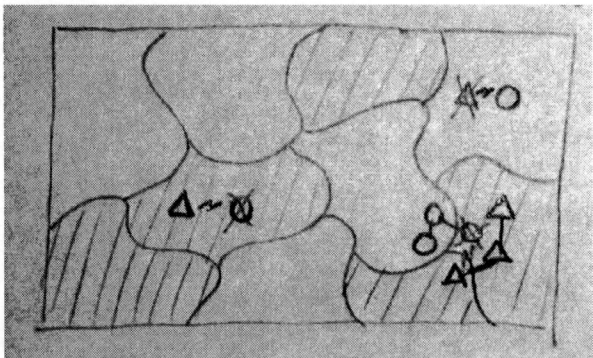

Figure 6.4 Storyboard: A one-to-one battle of two adversaries is won by the player who is in her/his favourable area. The playing field is divided in 50% red and 50% green areas. So, for example, if you are red you win in a red area and lose in a green area. In a confrontation between teams, the players belonging to the larger team win.

The latter idea (see figure 6.4) emerged as the most interesting, but we decided to keep the game logic as simple as possible for our first prototype, to ensure technological efficiency. GPS inaccuracy and network lag would not make the experience of zones seamless, so we chose the first solution above instead. We expected that the gameplay would not rely so much on computer skills (i.e. pressing the 'tag' button faster), but rather on the efficiency of the wireless network, on which we had little control. Depending on which player's 'tag' event arrived first, the server would determine the tagging event.

6.2.2   Group formations and swarming

From the very beginning we were thinking how we could introduce team formation and other collaborative mechanisms to encourage emergent collective behaviours among groups of people and enhance the social experience of the game. Figures 6.5 and 6.6 show early storyboards, inspired by the linking hands variant of tag mentioned above: linking with members of the same team as a means to become stronger.

We wanted to explore dynamic group formations, collective activity and movement in the city. Particular questions we considered were: Would linking be avoided or sought for? How large would the groups be? Would they become evident to others? Would people be able to keep playing, tagging, interacting with others and their environment etc and still maintain proximity with their group (figure 6.7)? How would these groups move in urban space and how would people communicate/keep awareness of each other over distance? Would we observe flocking and swarms between opposing groups?

Another idea was to reward players for the amount of time they spend linked with others was for them to get 'group power' based on the number of their links and how long they would maintain them (figure 6.8). This would then make them stronger against other people's tag. Linking could provide ground for a variety of social behaviours and interactions, both at individual and group level. Also, less active players would be accommodated in the game as well, because more active players would be motivated to link with them and keep them in their groups, to have stronger teams. Then the 'weak links' would need to try to follow along, so they would have a motive to get engaged. We imagined that people would undertake various roles, like 'lone chaser', 'runaway', 'team player', 'organiser', even 'spy' etc. Interesting situations could emerge by experimenting with the basic elements of the game. For example, if a player tagged an opponent and then tried to link up with him/her, would the opponent accept it? What could happen if tagged players started linking to each other?

While brainstorming about linking mechanisms, several issues emerged. For example, how would friends, who want to find each other and link, *communicate over distance*? One suggestion was for players to select a target location and send it to other team mates or a pre-specified buddy list. Also, preformatted messages could be used for quick communication. More detailed designs would need to be developed, because representing efficiently direction, the flow of movement of other people while being on the move is a particularly challenging information design problem, especially since the small display would need to be updated all the time.

Figure 6.5 Storyboard: creating a live social network. By finding peers in the vicinity players can link and both become stronger. (NB: the tapping on the screen is only a suggested interaction, for example people could also link via Bluetooth or infrared.). If linked team mates move too far apart from each other, their link breaks.

6.2.3 Views

We also considered having different views:

a) A local view, showing other players in vicinity. For this, we favoured an *implicit* view (figure 6.9), for example a radar display, which conveys proximity rather than actual location, as in a map view. This follows our principle of lightweight design, to facilitate interaction with the surrounding environment with appropriate presence information overlay, instead of providing detailed displays and navigation aids that would draw all the focus on the screen.

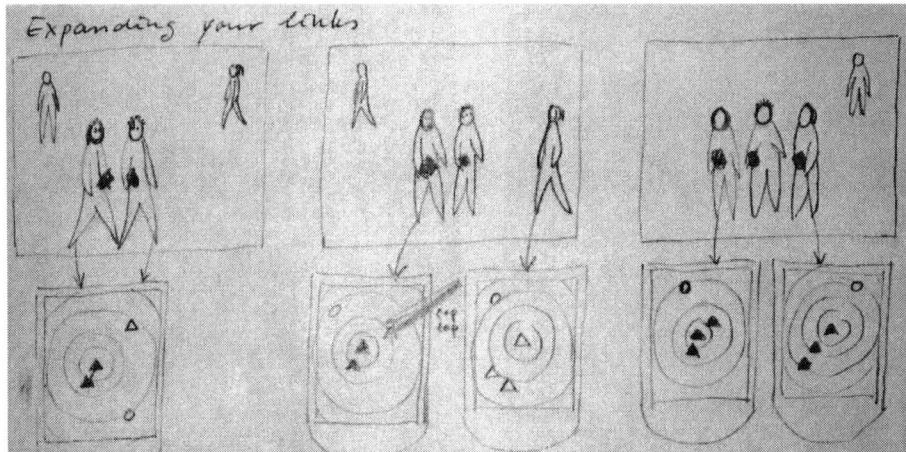

Figure 6.6 Storyboard: expanding the social network by adding more links to the group, each person can do that on their device.

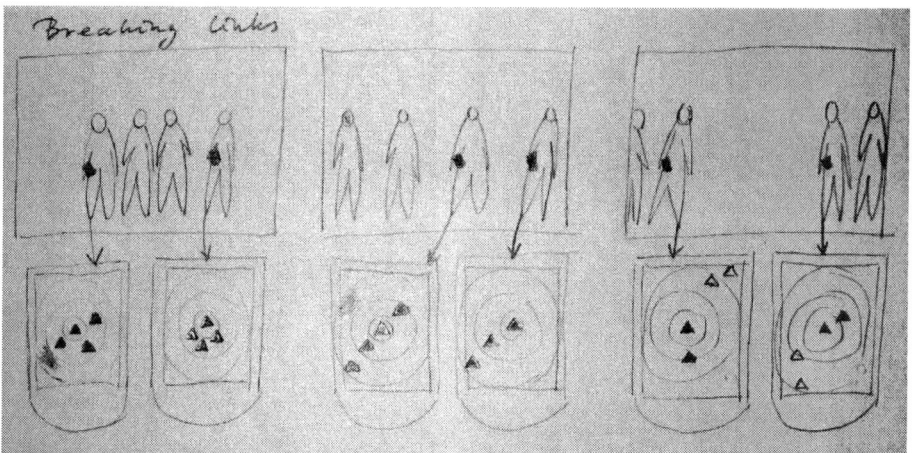

Figure 6.7 Storyboard: moving around without losing links. A player does not need to stay close to all of their links. One player can act as bridge between others. But the chain would break into 2 pieces for example, if two people in the middle move too far from each other.

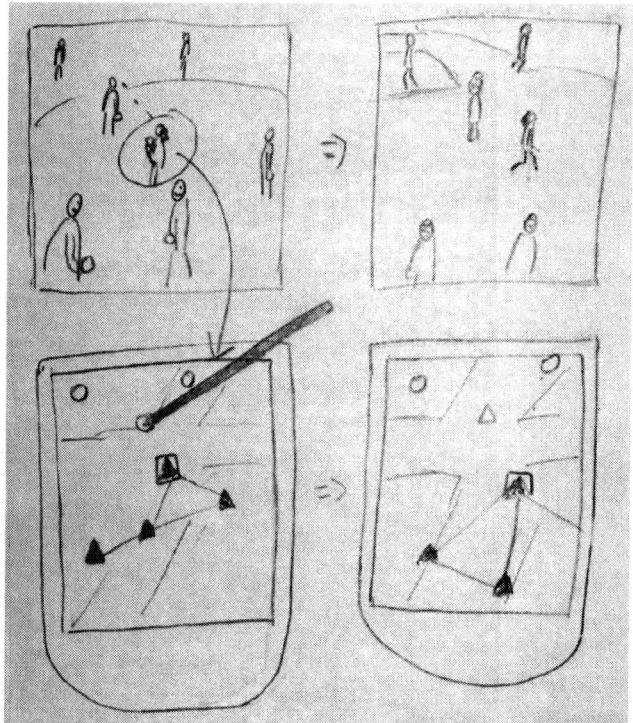

Figure 6.8 Storyboard: the most 'networked' one wins! Consider the scenario in which you are a member of a team of 4 and you find someone to tag, without knowing how strong she is. You tag her. She has 2 links while you have 3 at the moment, so she loses. Her links break as soon as she becomes tagged, but her other two ex-team mates are still linked with each other

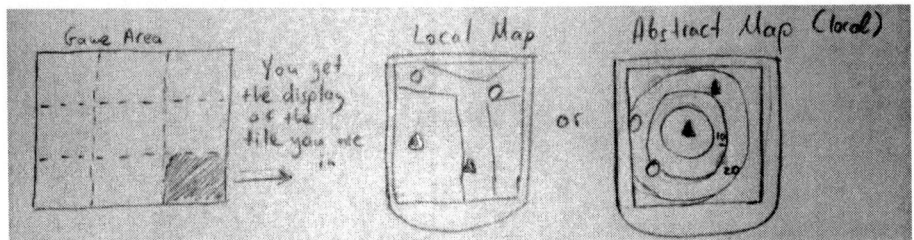

Figure 6.9 The playground is divided into sections, say for instance 9. The device displays the map of the tile the player is located. This local view can be either explicit, a street map placing players to particular locations, or, better, implicit, a radar indicating how close are other people to the player.

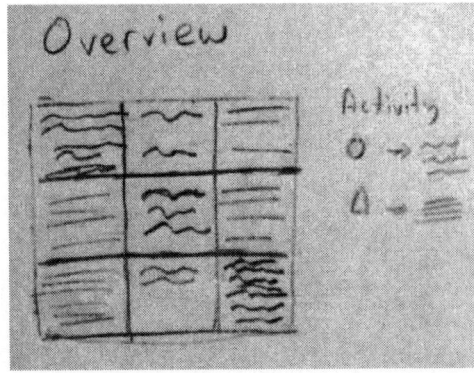

Figure 6.10 The overview would be purposefully abstract, to give an impression of the presence of other people in different areas of the playground and how active they are (tagging & linking). Here, each pattern represents an area where one of the two groups has greater presence. Pattern intensity indicates how much activity is happening.

b) An overview, showing crowd presence in the whole game arena, which could range from the size of a square to a whole city. This view would need to be even more abstract and implicit in order to be scalable, for example blurred colours, clusters or coloured tiles representing the majority of team members present in an area (figure 6.10).

6.2.4   Metaphor

Another design problem was finding an appropriate *metaphor* for the two opposing teams, to spark players' imagination. Tagging as in a children's playground can be quite abstract in the adult world, so for this reason we decided to provide two identities, to retain people's interest in the game. Social psychology studies (Tajfel 1970) confirm that people tend to associate with even minimal group identities and feel part of that group. The metaphor should help people to develop their own ideas about what they can do in the game and encourage unpredictable behaviour to emerge. Metaphors can also close down certain routes for players' imagination, therefore they should not be too strict and all encompassing. Our metaphor for the two teams needed to be in accordance to our principle of design for emergence, therefore:

a) It should *make sense* to people, to motivate them and encourage them to *extend it*. The metaphor should draw on real life experiences, to allow the users to relate to it.

b) Identities should be *persistent*. Our experiments with bumper cars in chapter 5 showed that if the group identity is constantly changing, people lose interest and often end up playing individually. With persistency we expect to have emergent interactions between the two groups: growing in members, flocking, etc.

c) The *two identities must be equally valued*, not biased by meaning. So there should be no particular advantage or political association for belonging to one or the other.

d) The metaphor should not get in the way of the *idea of tagging* – this is what the fun is all about, the simple idea of finding someone and changing their state, not engaging in battle or solving a problem.

We chose a simple and abstract metaphor for the prototype: Red versus Green. Colours are often associated with team membership in sports. We wanted to see whether this metaphor would be sufficient to generate a feeling of 'team belongingness' and participation in our game. This colour-based metaphor satisfied the minimal design principles and could also be extended in an urban context, for example by suggesting an association with pedestrian traffic lights. In this way, objects in the physical environment could be used in the game in creative ways: we could designate the traffic lights as a place where one can switch identity. So, if a player got tagged and wanted to get back to his or her own team, he or she could stay for half a minute or so in this area and be switched automatically.

Finally, we also considered attaching more meaning to the action of tagging, such as getting a 'trophy': imagine that when you tag another player, you also get something personal from him/her, a photograph or some information about him/her. This would give rise to all kinds of talk and rumour and enhance the social aspect of the game.

Figure 6.11 Storyboard: surprise someone with your peers. A player indicates to her team mates (links) a target person to surround without getting noticed. As soon as the unsuspecting opponent gets surrounded by the 3 players, he becomes tagged immediately!

In order to encourage group formation and flocking we considered the idea of surrounding an opponent to tag him or her automatically (figure 6.11). Although this was not implemented as part of the final prototype, as we will see in section 6.5 on findings it became a reality by the players themselves surrounding lone opponents with their team mates spontaneously!

This was not the only case of a behaviour we thought during the design process which was actually performed by the players, without the need of any functionality to support it in the prototype. Another example was the idea of going under cover or becoming a spy to tag opponents secretly, which we considered in the design process as an extra feature for players to discover through play. As we will see in 6.5, we observed a spontaneous 'secret assassin' who performed precisely this role without it being part of the design or being mentioned at all.

6.3 The CitiTag game

6.3.1 Limitations and design considerations

When designing the final prototype there were many factors to consider, such as the technological constraints and limitations of carrying out this research in the form of organized user trials, where everything needs to be coordinated to perfection to achieve success. The prototype was implemented in collaboration with HP Labs Bristol and used their Mobile Bristol authoring tools for location-based mediated experiences. Using Mobile Bristol posed some constraints: we had to use GPS for location positioning and Wi-Fi for network communication. GPS accuracy depends on certain conditions, such as the weather and the availability of satellites. We could not have a compass showing the orientation on the interface for this prototype, but we could design the interface in Macromedia Flash, which gave more flexibility and opportunities for creativity.

We also had to decide on a convenient city location for the trial, a central place in Bristol where we would have good GPS and Wi-Fi signals, but which would also be appropriate for the type of game we have been designing, with enough people around and physical objects to facilitate emergent play. We also needed to be able to easily observe what would be happening from a distance. For the purposes of our trials we had to limit the number of participants to a maximum of 20 people because we had a limited number of devices to lend and it would also be really hard to organize these events with a larger number of simultaneous participants. Other design considerations were visibility, interface legibility in daylight, device specific input and output (using iPaq Pocket PCs), battery life etc.

We then stripped down all storyboards to the simplest combination possible to implement for user trials, which would act as proof of concept. The idea was to start as simple as possible, in accordance to our design principles, and to build up in the future on the existing game skeleton. The simplest interaction within our game was *tagging* and *untagging*. The final prototype is described in the following section.

6.3.2 CitiTag Design

Our final prototype, CitiTag is a multiplayer, wireless location-based game, played using GPS (Global Positioning System) and handheld, iPaq PocketPCs connected to a wireless network. As a player of CitiTag, you belong to either of two teams (Reds or Greens) and you roam the city, trying to find players from the opposite team to 'tag'. When you get close to someone from the opposite team, you get the opportunity to 'tag' them: an alert appears on the screen with a sound (figure 6.12 a). You tap on the screen with your thumb to 'tag' the other person.

You can also get 'tagged' if someone from the opposite team gets close to you and 'tags' you first (figure 6.12 c). If this happens, you need to try and find a team member in vicinity to set you free, to 'untag' you. Similarly to tagging, players receive an alert whenever a 'tagged' team member is nearby that they can rescue (figure 6.12 b), provided there is no opponent nearby to tag them at the same time. There is a 'group state' feature, so players can see how many people are free and how many have been tagged in each team at any time (see top of screens in figure 6.12). Players see how many people they tagged and rescued individually at the end of the game. The game logic is simple, based on all possible interactions between four players, two from each team, as described in the flowchart in figure 6.13.

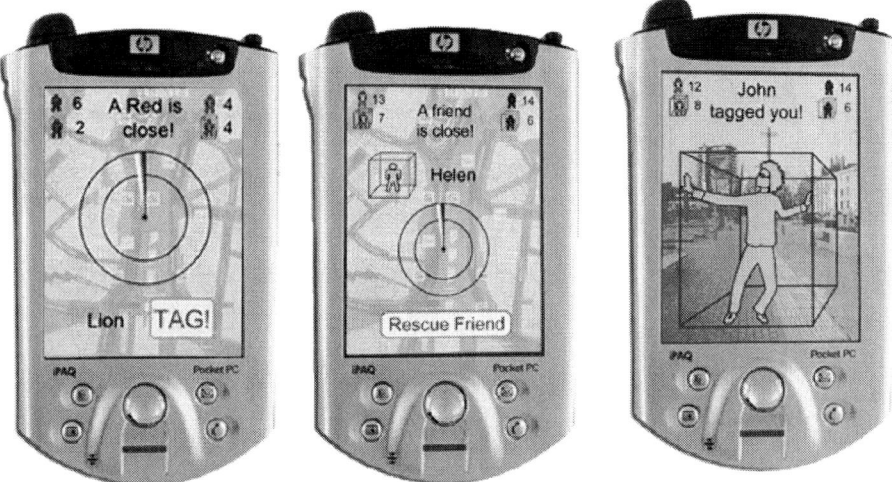

Figure 6.12 CitiTag screens: a) A tag event appearing in an encountered between two members of the opposite teams b) A prompt to untag, to rescue a tagged team member c) The user got tagged

CitiTag is our interpretation of playground 'tag' as an everyday, mobile technology mediated experience that is appropriate for adults. It combines effectively several aspects of our initial mixed reality concepts in chapter 4 (PixelTag, TravelGame, Party activities) all in one. CitiTag is better than PixelTag in that it is truly ubiquitous, it superimposes virtual presence on physical, drawing attention to the real world and it capitalises on proximity to create a mixed reality experience. It is an outdoors, urban game like the TravelGame, but it is much more lightweight, because it is designed to support awareness of the surrounding environment and facilitate interaction with other players. CitiTag is a social game and depends on the presence of other people, like the party activities in chapter 4: the more people are participating the more interesting and variable the emergent interactions are likely to be. Also, it is not limited to the context of a particular event, but can be played in various moments of daily life and urban locations.

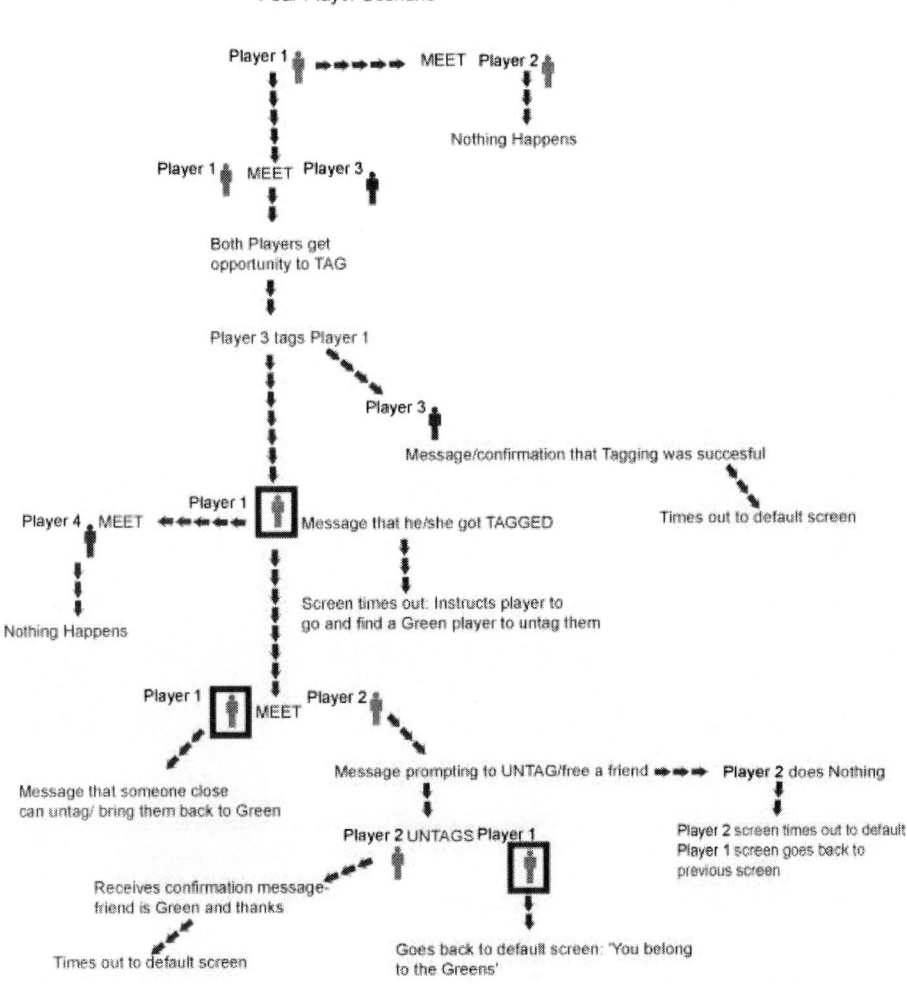

Figure 6.13 CitiTag flowchart: we can see all the game possibilities in a four player scenario. A minimum of four players is required, two of each team to explore all the game branches, both tagging and untagging.

CitiTag has been designed in accordance with our four key design principles presented in chapter 3:

1. *Symbolic presence.* The game events are communicated through simple presence states and alerts along the lines of 'I am Green and tagged' and 'A Red is close' respectively. Virtual presence is superimposed on physical and players are challenged to identify individuals in the real world.
2. *Lightweight design.* We are motivated by the hypothesis that very simple game rules based on symbolic presence states can result in an enjoyable social experience, stimulated by real world interaction among players. Therefore, presence is communicated in an abstract way in the prototype, indicating proximity rather than a person's actual location. The players' true location is not revealed and they receive only the name or nickname of a person being

really close, whom they can 'tag' or rescue. This also helps to discourage people from looking at the screen too much to pinpoint others' exact location. Instead it serves as an alert aiming to draw attention to the surrounding environment, supporting awareness of other players, and to give a clue for players to try to identify the person appearing on their screen.
3. *Potential large scale.* Although we tried our game with a limited number of users, it is potentially scalable, if made available for mobile phones for example, because of the simplicity of its structure and game rules. This game could potentially evolve to an everyday experience with a mobile phone one could sense in the future, while walking about in a city centre.
4. *Design for emergence.* The familiar metaphor of tagging originating from the famous children's game and the idea of the two teams are very simple, to allow people to extend and bring new meanings to the game as they discover their own cooperation strategies. Players can also reveal or hide their presence, both virtually (e.g. stay under a building or covered area to lose GPS satellites) and physically (e.g. hide behind a tree). By hiding or revealing presence at appropriate moments, participating in spontaneous 'playground' play can become in the future an in-and-out experience, part of the fabric of our everyday life. CitiTag allows space for people to experiment within the context of the game and it is designed to encourage spontaneous interactions among people in a city environment like running, forming groups, hiding etc.

Certain behaviours we expected to emerge. Other behaviours would come as a surprise. We also wanted to see how players' behaviour affected the behaviour of people who did not participate, but just happened to be at the same location during the game and the other way around.

6.3.3 CitiTag system architecture

The diagram in figure 6.14 shows the system architecture for CitiTag (a tech report for Macromedia Flash developers by Quick and Vogiazou, 2004 describes the components with more detail), which is part of the Mobile Bristol project, by HP Labs Bristol (Hull et al, 2004). The Mobile Bristol Application (Hull, 2002) is used to read the location data from the GPS receiver and to pass it via an XML socket, to the Flash Game Client, the program that is used to actually play the game on a Hewlett-Packard, iPaq PocketPC. This program is written in Macromedia Flash and it also connects, wirelessly, to the CitiTag application on the Flash Communication Server (Macromedia, 2004). The server incorporates the game logic and sends all connected game clients updates about their state (tagged or free) and nearby devices they can interact with (tag or untag). There is a browser-based Administration client also written in Macromedia Flash which is used to administer and monitor the game.

CitiTag is different from all other Mobile Bristol projects in that it has these multiplayer functionalities, users interact with each other through the game. Technically, it uses both GPS and Wi-Fi, whereas the other projects are based on either one or the other. This technological combination is complex and has an inherent contradiction: Wi-Fi works best in small, defined spaces, while GPS works best in larger, open spaces. Thus, having both working in full together has been quite a challenge in the development of CitiTag.

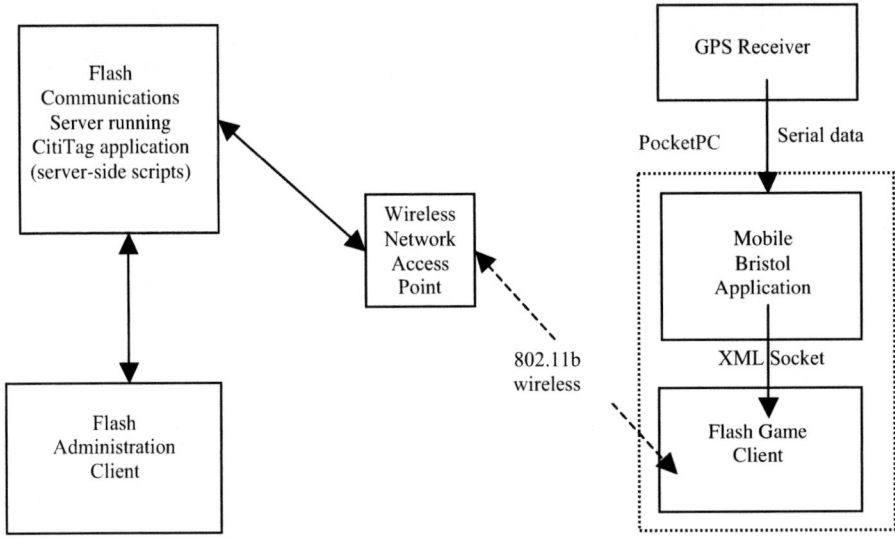

Figure 6.14. CitiTag architecture

6.4 Method: user studies

We used a combination of qualitative and quantitative research methods in our empirical studies. Two user trials were carried out: a pilot trial with 9 participants in an open field space at the Open University campus in Milton Keynes and a trial with 16 participants in a square in the city centre of Bristol. Participants played the game outdoors for approximately 40 minutes with intervals for restart, debugging or quick briefing when necessary. There was a briefing session of 15 minutes before the start of each trial when participants received the kit and instructions on how to use it and how to play the game. Trying the game in two very different locations allowed us to compare the emerging interactions in the context of the particular environment where the trial took place. We took video of both trials, aiming to capture the experience of game play, as opposed to simply registering what was happening. Therefore the cameraman (Bas Raijmakers) walked or even ran with the players. This allowed us to observe what people did, how they reacted to events in the game, see their face expressions and hear what they were saying to others. Also the close presence of the cameraman often elicited them to comment on the game. On such occasions they freely ventilated their thoughts, very much like protocol analysis where users are asked to 'think aloud' while solving problems or performing a task (Ericsson and Simon, 1996). The material was edited down by Bas Raijmakers to 3 minute films for each of the two trials. These films allowed us to convey the game experience and show some crucial observations to others. The user trials were followed by group interviews, in addition to which all participants completed a questionnaire about their experience with CitiTag. The group interviews lasted 30-45 minutes and were open discussions loosely structured around the main research themes: experience of gameplay, the game as part of everyday life, group cooperation and strategies, awareness of others and interaction with the device. These discussions were also recorded on video for further analysis. The questionnaire was designed to collect more detailed data on the research themes as well as the usability of the device and game interface. The questionnaire consisted of graphic rating scale questions (Stone, Sidel, Oliver, Woolsey and Singleton 1974) interspersed with open questions. Participants were asked to put a mark on a 100 millimetre line between two extremes, the extreme left being 0 millimetres ('Not at all' in many questions, scored as 0) and the extreme right 100 millimetres ('Very Much', scored as 100). The statistical results reported in the Findings section below concern correlations between two variables, denoted by the letter 'r', accompanied by the relevant degrees of freedom ('df') and the likelihood ('p'), indicating how much the results could have come about by chance, with 'p' being significant when $p < .05$.

Participants were recruited via email sent to student and departmental mailing lists. The invitation was open to anyone and although we aimed for a student population, the majority of our volunteers (ten out of sixteen) in Bristol were HP Labs employees, eight of them decided to try the game as an exercise for team building. This brought an interesting dimension to our research and we tried two different variations of the game: one where team members were not known and part of the challenge was to find out who was on which team and one where the group was divided into two visible teams, with the team practising team building starting together on one side of the square. All participants in Bristol used nicknames in the game. In our first trial in Milton Keynes we tried two different variations of the game: one with participants' real names and another one with nicknames.

Organizing both of our experiments involved a great effort and detail as everything needed to be coordinated to perfection for sixteen people to be able to play together simultaneously outdoors. For instance, in Bristol, a support 'crew' of ten people was necessary to prepare and run the trial. Our trials were not without problems, in particular, having a stable wi-fi signal and accurate GPS positioning at the same time on each device was a great technical challenge which we dealt with in the best way possible. Planning these experiments in the form of events was necessary to achieve a critical mass of players and test the concept at an appropriate city location.

6.5 Findings

We present the results from our two user trials and discuss their differences and similarities on fronts of interest: emergence, game experience, awareness and team 'belongingness' and design.

6.5.1 Game Experience

At the Open University trial all nine participants enjoyed playing the game a lot, with a mean of 82.1 on a graphic scale of 100 millimeters rating enjoyment from 0 (not at all) to 100 (very much), with 6 out of 9 ratings being between 80-95. Four participants mostly enjoyed the competitive aspect of tagging, which two of them identified as 'hunting instincts'. Two participants liked that CitiTag is an outdoors game and the rest three participants enjoyed: the ease of play, not knowing 'who is who' in the game and the fact that it reminded them of 'cops and thieves' game from childhood. In response to the question whether they would like to play this game more than once, all participants gave a positive response. They mentioned that 'they hadn't explored the full potential of this game yet', 'there can be many game scenarios' and that 'it is different every time you play it'. One participant commented that 'it is nice to belong to a team' and that 'it is even more interesting when you don't know who is in your team and who is an opponent'. Six participants said in the questionnaire that what they least enjoyed was dealing with technical problems, GPS errors and wi-fi drop outs.

During the game participants felt very engaged (mean 79.2), quite amused (mean 69.8) and very social (mean 92.0). They generally felt they were in control of what was happening (on top of things, mean 85.0) and any confusion they felt was associated with technical failures, as it became evident from their comments both in the questionnaires as well as during the group interview. We asked participants also to rate how tense or relaxed they felt during the game and received variable responses ranging from very tense (minimum 9.00) to very relaxed (maximum 93.00), which indicates that the experience of play is personal and can vary a lot.

In our Bristol trial on the other hand, the sixteen participants fairly enjoyed playing the game with a mean of 49.25. Three people said they would definitely want to play it more than once and seven more said they would like to play again an improved version with the technical problems resolved. When asked what they most liked about the game six participants said they enjoyed playing it outdoors, three said they liked the technology, two liked working out who is who and one said he liked to play tactically, with known team members. Other aspects they enjoyed were 'playing 'secret' in public', the 'sense of freedom' and 'social interaction'. Most participants' responses on what they least liked about the game (eleven out of sixteen) focused on technical problems.

Our participants in Bristol were quite engaged when they played the game (mean 60.12), quite relaxed (mean 65.37), fairly amused (mean 53.6) and they felt moderately on top of things (mean 49.43).

Bristol participants felt reasonably social (mean 54.93) with 10 out of 16 people rating between 60-85. This result is however significantly lower than the result from the pilot study (mean 92.0) ($F(1,23)= 9.336$, $p<.01$). Our participants also felt moderately isolated (mean 54.62) with 7 out of 16 people ranging between 40-60. This result is significantly higher than the OU study result, where participants felt less isolated (mean 35.6) ($F(1,23)= 4.452$, $p<.05$). Participants' overall enjoyment of the game (mean 49.6) was significantly lower than that of the OU trial (mean 82.1) ($F(1,23)= 12.556$, $p<.005$).

These are the reasons we believe why the first experiment was more enjoyed than the second and participants felt more social:

a) The Bristol group was less consistent than the user group at the OU; at the OU all participants knew each other and interacted a lot during the trial, even when technology failed they still made fun of the whole experience. Being in a more confined and exposed location made it easy for them to communicate with each other over distance and there was always someone near they could go to. In the second experiment half of our participants were from the same team but the rest were fairly mixed and they did not interact that much from the very beginning. Frustration with technology was a significant factor, some people felt isolated because they were out of the game due to technical problems. For example, one participant in Bristol was most unfortunate of the whole group to have constant technical problems, even though he tried different devices, so he became very frustrated and also felt most isolated (97) of all.

b) The second reason is that the largest proportion of the Bristol group had a technical professional background (ten out of sixteen) and as they were losing control over the technology, this caused feelings of confusion and hampered the experience. Dealing with the ambiguity of an outdoors environment, where even a bus passing can cause wi-fi signal drop out can be very frustrating.

c) Possibly the expectations of the two user groups prior participating in the trial were different too. In particular, the OU user group knew that it was hard to run this trial because an earlier trial had to be cancelled because of technical problems, while the user group in Bristol knew nothing about the prototype aside from the brief description in the invitation they had received.

We have identified several factors that enhance or hamper the experience of playing CitiTag and we discuss them below:

1) Location.

Our user trials revealed that the experience of playing CitiTag can vary depending on the location where the game is being played. We noticed through observation significant difference between the quick, action game sessions in the OU trial and the more strategic ones, with evolving cooperation and group convergence in the city environment in the Bristol trial. Players were able to move peripherally, hide and work out strategies with others in the city, actions that were much more difficult to perform at the Open University trial, because the field area was small and completely exposed. At the OU trial, our participants pointed out that they were not really aware of what else was going on around them beyond the game (their mean rating of general awareness was 29.11 on a scale between 0 (not at all aware) to 100 (very aware). This is almost self-evident, there were no passing vehicles, not that many people around and the space was defined, so there was no need to pay attention to anything else beyond the game (e.g. passers by, traffic lights) and there were no other attention stimuli beyond the trial activity. On the other hand, in the city environment of the Bristol trial, our participants were quite aware of what was happening in St. Augustine's Parade beyond the game with a mean 48 and with 10 out of 16 people ranging between 40-90.

In the OU trial participants did not find the location particularly suitable for this game (mean score was 54.2). One person who gave a high rating (89.0) also gave a contradictory comment to 'play the game at a location where not everybody is continuously exposed in an open field'. The person with the highest rating (93.0) said that with extra people you wouldn't know who is playing and who is not. Other players however, are likely to enjoy this kind of uncertainty as indicated through their comments. Eight out of nine participants agreed during the interview that the location was not particularly suitable and explained that they felt too exposed and they needed more space, obstacles, players and other people around. They wished to explore further the possibilities of play, collaboration and strategies in the city.

In Bristol, however, the location did encourage the emergence of collaborative strategies as part of the game. In response to the question on the suitability of location most questionnaire responses (11 out of sixteen) ranged between 40-80, with average

55.12. Eight people who were positive about the location (they rated its suitability above 60) thought it was good to have other people and objects around. Two people felt too exposed and wanted more places to hide and three people found the location too busy and noisy. One participant liked the fact that he had to study the environment. During the group interview, participants gave positive comments when discussing the location and one participant suggested other city areas which would be more convenient, such as pedestrianised areas with no traffic and less noise, and areas *"with lots of concrete blocks you can hide"*.

Playing CitiTag in a public space, rather than in a game-specific location (e.g. like paintball) is important and a big part of the game's identity, as one participant in Bristol mentioned: *"I was trying to compare it with other 'tag' types of games, if you think of laserquest, where you go into an environment of dark lights and people wearing suits, walking around with guns, but here it was a form of tag game which actually you **could** play and it wasn't awkward to play in public"*. We therefore have concluded that an urban environment is more appropriate for this type of game.

2) Group dynamics and the social aspect.

People also make the experience different every time, individual behaviours emerge and other players respond accordingly, group dynamics constantly defining gameplay. A great part of the overall experience is the social aspect and strategy in the game. One participant at the OU commented that *'it is nice to belong to a team'* and that *'it is even more interesting when you don't know who is in your team and who is an opponent'*. Another participant said that *'it is different every time you play it!'*. OU trial participants said in the group interview that they preferred playing with nicknames rather than with their real names, because as they all knew each other the gameplay was too easy and finished quickly, whereas it became more intriguing with nicknames, when they were trying to figure out who was who. In Bristol, some participants said in the group interview that they liked trying to find out who was part of which team and other participants expressed a preference for playing with a known team. This suggests that the game has two different yet complementary dimensions: an initial phase of exploration where the player is alone, trying to find team players and tag opponents, much like a spy, and then a second phase when the team is known and people cooperate against the opposing team. When analysing both experiments, we found a significant correlation between enjoyment and feeling social across both experiments ($r= .542$, $df= 24$, $p<.01$). This shows that in our game, the social aspect is particularly strong in the overall experience.

Our participants in Bristol suggested some ways they would like to see CitiTag evolve into a more strategic game, leveraging social skills, for example though matching profiles, encouraging team formation based on the profile, more decision making and trade-offs, like being able to choose the priority between tagging and rescuing. The element of getting to know your opponents or friends was flagged as particularly interesting to explore through this kind of game, which can be combined nicely with socializing and meeting people.

3) The real world as an interface.

Our results suggest that immersion in CitiTag is associated with the real world and people around. In particular, in the OU trial we found a significant correlation between 'enjoyment' and 'usefulness of sound alerts' ($r = .979$, $df = 8$, $p <.001$), so the more a player enjoyed the game, the more useful they found the sound alerts (figure 6.15). The use of audio was appreciated by participants in both trials (mean 76.3 in OU and 62.93 in Bristol). Audio cues were good for awareness, because they often liberated people from looking at the screen too much, which took away attention from the actual game interface – the 'real' world. In Bristol, one participant gave as an example of the kind of interaction we should be aiming for through the use of audio: the scary sound in the film *Aliens*. This sound warned that something was coming close and was becoming louder and more regular as the 'alien' was approaching, so as you could not see anything, it was very scary just with the sound. This example highlights the potential of sound as an awareness feature, enhancing the player experience.

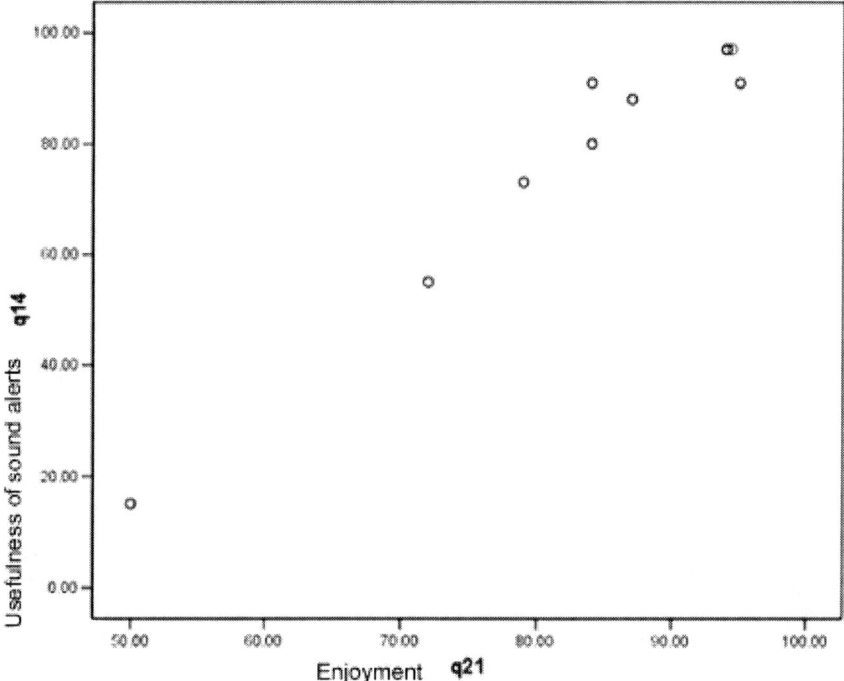

Figure 6.15 Correlation between 'enjoyment' and 'usefulness of sound alerts' at the OU trial

We found a significant correlation in the Bristol trial (figure 6.16) between awareness of other players and enjoyment (r= .694, df= 14, p<.005). This result indicates that participants who were more aware of other players, enjoyed the game more than others who, for instance, focused on the screen, either trying to solve technical problems or figuring out what was happening, rather than observing the people around them. For example, three players who did not look as much at the screen as others (their ratings on the 100mm scale were 68, 46 and 66) were very aware of other players (85, 99 and 91 respectively). They also enjoyed the game (75, 74 and 60 respectively) and mainly commented on the technical difficulties with GPS and wi-fi as negative aspects in their experience.

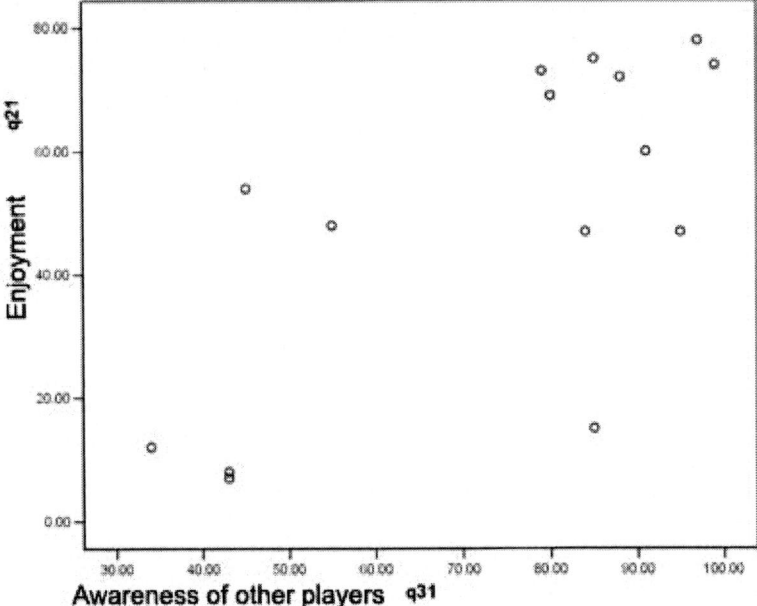

Figure 6.16 Correlation between 'enjoyment' and 'awareness of other players' at the Bristol trial

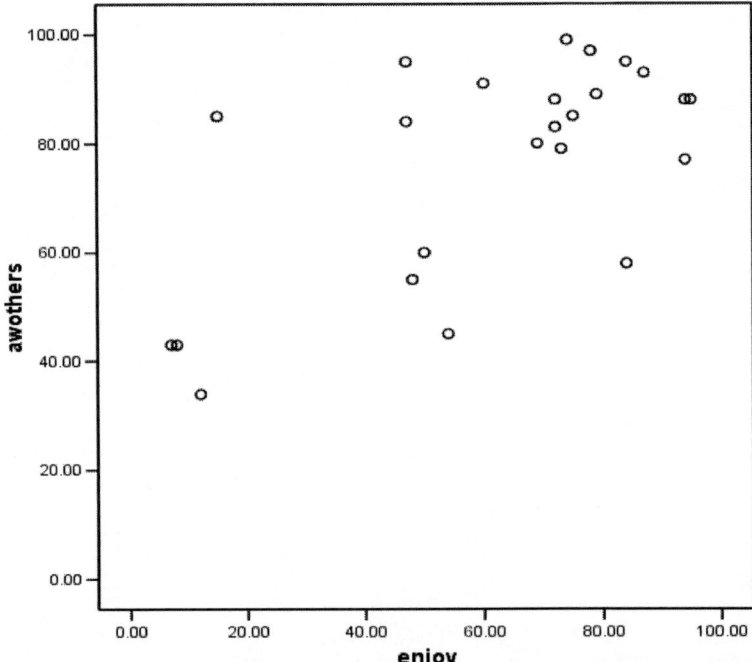

Figure 6.17 Correlation between 'enjoyment' and 'awareness of other players' across both trials

Similarly, a significant correlation between awareness of other players and enjoyment was also found across both the OU and Bristol experiments (r= .640, df= 23, p<.005) which further indicate that immersion in the physical and social environment is more enjoyed (figure 6.17), so the game and device should promote and encourage this kind of interaction. These results confirm our lightweight design principle, discussed in chapter 3 of this thesis, as the right approach for this type of mixed reality (physical and virtual) social game.

4) The mixed reality misconception.

Observation and participant feedback in the interviews revealed a factor that decreases enjoyment and hampers the game experience: the mismatch between reality and what a player expects to see on screen. In the group interviews participants said that they expected what they see in the real world to be reflected in their device with a relevant timely alert. This was not always the case with GPS errors and temporary losses of wi-fi outdoors, causing frustration. This is further confirmed from the video footage: a participant looked at his device in anticipation for an untag event to appear while standing next to a member of his team and became increasingly frustrated because nothing was happening. Technical problems sometimes caused strategies and attempts for collaboration to fail, which caused even more frustration. Also, identifying participants by their nickname and working out who was who became a difficult process. At the OU trial, participants did not find it particularly easy (mean 45.4 with most ratings between 20-65, from 0 (not at all easy) to 100 (very easy) to identify a person in the physical space by their nickname whenever they received a tag/untag alert. In Bristol, most participants found it very difficult (mean 26.2 on ease with 11 answers ranging between 0-30), also because they were more in number and more dispersed. When a few people were together in a small area, it was hard to work out who was who and this was made even harder by GPS errors and delayed alerts. We think that not trusting the accuracy of the displayed information was a major factor decreasing the enjoyment of the whole experience.

These observations made us think towards other design directions, the need to rely less on accuracy and to employ a greater level of abstraction whenever possible. So by having less fine grained proximity, the virtual state (e.g. a player of the opposite team is close) becomes more ambiguous and any occurring errors would be less obvious, therefore more acceptable. We discuss this approach further in the next chapter.

Our results here suggest that playing CitiTag is an engaging social experience on the 'real world interface', which can vary as it is stirred by group dynamics and interaction among players and it is dependent on factors like location and the match between virtual and physical reality.

6.5.2 Emergence

Observation during the trials and video analysis revealed a range of emergent behaviours, both individual and collective, which show how our participants explored the game, technology and the social context by pushing the limits of what was available to them. In the Bristol trial, this process became clear from the log of the game: initially one team quickly won the other, but later, as strategies evolved, the teams learned to fight back, so the game lasted longer and the numbers of tagged people kept going up and down with often surprising last minute victories. Similarly, in the trial at the Open University, once we introduced nicknames and the game became more ambiguous, participants spent more time trying to find out who was part of which team, so the game lasted longer and also became more interesting for them.

Our observation is that emergence occurred when people pushed the boundaries of both the virtual and physical world. Here are some illuminating examples:

1) Pushing the technology

In both trials we observed participants really exploring the technology and finding what works and what doesn't, so in the Open University for example, we had two people who tried to estimate the range for the tagging event by getting close to each other slowly and moving further away, watching when the tag event would appear. Often people took the role of the 'amateur scientist' in both trials, looking at each other's screens (figure 6.18), trying to solve occurring technical problems with GPS and wi-fi and discussing how the technology worked. One participant in the Open University started running around trying to pick up any signal. What is most interesting is how he exploited the technology: when he would see that he had a good signal and everything was working well, he would try to get close to as many people as possible (figure 6.19).

Figure 6.18 Players look at each other's screens to solve technical problems and figure out how the technology worked.

Figure 6.19 A player runs around in the OU trial, trying to pick up a signal with his device.

Figure 6.20 Players team up in a pair to be able to rescue each other.

In the Bristol trial, we often observed participants teaming up and moving around in pairs (figure 6.20). Six participants mentioned in the questionnaire the pairing as a strategy they used in the game. In this way they were in a more advantaged position than a lonely opponent: even if he or she tagged one of them, the other could try tagging the opponent and then rescuing the tagged team member. This tactic often resulted in clusters of team players moving along together in a group. One participant in Bristol commented during the group interview:

"*It seems to work! We got about five people in one session and kept ourselves released and then kept walking and we got more people, so it did seem to work*".

Our conclusion here is that we would not need very sophisticated technology (e.g. maps, accurate location, team formation mechanism etc) to encourage people to converge in groups and move along together, but that such interaction can emerge through the exploration of simply designed technology, based on two actions: tagging and untagging.

2) Pushing the metaphor

The simple tag metaphor helped people relate to the game and use their imagination to extend its meaning through their interaction. In the Open University trial, one participant kept his hand up to indicate he was tagged in order to attract attention, so that someone would come and untag him (figure 6.21). He said that he remembered his classmates did this in 'tag' games in childhood. The same participant also went and sat in a corner waiting for people to come by and untag him and in fact, other participants approached him. In Bristol, waving or raising hand when tagged to attract attention, was a typical behaviour, performed by at least five different people. This strategy however, did not work as expected, because it also attracted the attention of members of the opposite team who would sneak in to tag the potential untaggers.

Figure 6.21 Tagged players raise their hand to indicate they want team members to rescue them.

As one participant commented: *"You realised it gave the opposition a sign that you are heading there"*. Another participant mentioned it explicitly as a strategy attempted during gameplay: *"hang around tagged opponents to get the rescuers"*.

In the Open University trial, one participant admitted she kept pointing the device at other players when tagging them, even though she knew there was no need to do that. She described it as a kind of gut reaction. Gestures like this remind us other types of games (e.g. laser quest) and can be interpreted as a threat or sense of satisfaction (I tag you!).

Our conclusion is that people could relate to the 'tag' metaphor in different ways and often extend it. We had references to other playground games, too, for instance a couple of descriptions of what it felt like to play the game: one person said it was a kind of cyber 'cops and thieves' game and another that it reminded them of hide and seek, only that everybody hides and seeks same time. While physical behaviours associated with the classic tag game are not part of CitiTag's design, we observed that people could extend the metaphor of 'tag' by introducing this physicality in the game. Our conclusion is therefore that the use of metaphor is helpful in communicating a style of game 'It's like a game of tag', but then the actual tactics, behaviours and interaction techniques of the new game are defined through experimentation and trial.

3) Bending the game rules

Most games can have different levels of rules: rules that are enforced by the mechanics of the game, locally established rules and social and legal rules. For example if we consider the game of football there is a well established set of game rules. Local rules that interpret or bend those games rules can be agreed by the players and are enforced by a referee or the players themselves. Social rules such as not allowing ball games in certain areas restrict the venues and the areas that one can play. A parallel set of rules can be established for mixed reality games. In CitiTag game rules, enforced by the system, will not allow a person to tag others once he or she has been tagged. Other rules can emerge locally, depending on the context of the game. For example, in our trial at the Open University participants often asked explicitly what team others were in, making the game too obvious and easy, so we introduced a 'no speaking' rule. Social and legal rules exist in the environment, so it would not be socially acceptable to grab non participants and hide behind them or to damage or disturb features in the environment.

We observed several cases in which participants subverted the rules of play in creative ways. One participant in Bristol introduced a 'secret assassin' behaviour that no one expected, by breaking the team division game rule (figure 6.22). He logged on to the opposing team and when the game started he switched back to his original team whilst among opponents and tried to tag as many people as possible before getting tagged. This was very surprising because teams had been divided at that point and as the whole Green team was at the other side of the square where the trial took place, players of the Red team could not understand what 'hit' them right at the beginning of the game. So by breaking the rule, this participant also created an entirely new role in the game. We were fascinated by this emergent behaviour, especially because we had already considered a similar possibility for people to go under cover as a hidden 'extra' of the game design in our storyboarding process (see paragraph 6.2) and there it was, a player came up with this, without it being part of our design at all.

At the Open University, participants very quickly subverted the 'no speaking' rule by using gestures and body language to communicate what was happening, for example smiling at someone in a cunning way, trying to guess their team membership by exchanging facial expressions. One participant noted in the group interview an inherent contradiction in the game from the point we introduced the 'no talking' rule, between a) wanting to be *secretive* to obey the rules and b) wanting to be more *sociable* to have some fun. Local rules in mixed reality games could be negotiated and defined depending on the situation and the context of play (e.g. how exposed is the location, familiarity with other participants). Our observations revealed that bending the established rules can result in a creative activity (e.g. a new role in the game, new forms of interaction), therefore the design of such games should allow space for exploration and play with the rules themselves – a kind of transformative social play.

Figure 6.22 A 'secret assassin' tags puzzled members of the opposite team when they don't expect it.

4) Pushing the physical environment

At the OU trial, participants tended to converge altogether and to cluster on one spot in the field, coming from all sides. This was due to the openness of the space, which made everything immediately visible. Since there were no obstacles, other people or any other distraction, people immediately got closer to each other to have some action. In this way, the game finished really quickly and it was hard to develop strategies. As one participant commented: *"You couldn't stop clustering"*, even though this stopped the game. As they got more used to the game and after we introduced nicknames, they started approaching each other more cautiously while walking in the periphery, trying to figure out who was a friend and who a foe. This made the game more interesting. In Bristol on the other hand, people also converged and often moved along as a group, but unlike the OU trial where convergence resulted in the end of the game, team members converged to become stronger against the opposing team. Players in Bristol also tried to use the environment to their advantage by hiding behind obstacles when trying to follow secretly another person. Two participants said in the group interview that they tried to stay behind a bush for some time. However, hiding is not only physical as there is another form of hiding possible in CitiTag; one participant mentioned that if you go under the bus stop you would lose GPS so you could not be tracked any more, what we have identified as hiding in the virtual world, i.e. still visible by others, but not virtually 'there'.

It was also fun to observe the effect our user trial had in the social environment of the square in Bristol. People were curious about what was happening (figure 6.23) as they were watching many people with small back packs running or walking all around and some approached players and asked what was happening. Other, 'non-participants looked quizzically on', as one player reported. Interestingly, a couple of girls passing by, when they saw one of the players holding his hand up (to indicate that he was tagged), waved back and greeted the player in amusement.

It became clear after this trial that using the device and participating in the game automatically authorized people to be a child, but in a socially acceptable manner. In fact, people used the space as indeed their own playground, often running about, laughing and shouting over distance. It is important to note that our participants did not feel embarrassed to play this game in public, as the average score was 23.25 on a 100mm scale of 0 (not at all embarrassed) to 100 (very embarrassed), with 12 people out of 16 ranging between 0-40. In fact, one participant commented that playing it in public was actually a great part of it.

Figure 6.23 A passer-by has stopped to ask a trial participant about what she is doing and started a conversation.

Although our trial attracted attention, it did not have a disruptive impact on the social environment of the square because players respected the social rules we mentioned above, like not getting in the way of other people's activities. Skate boarders were also present in the same location, so the square was a public space allowing for play. Nowdays mobile phones are banned in many public spaces, because they are socially intrusive. The design of CitiTag aimed to facilitate peripheral awareness and we did not observe any cases in which engagement with the game and lack of awareness weakened the players' social responsibility, which is a common risk for mixed reality experiences.

5) Stretching limits of themselves and other players

In both trials, people pushed the limits of themselves by trying to work out strategies both individually and as a team. Frequently observed behaviours were running, often contagious, introduced by one or two people and followed by others starting running too (figure 6.24). They did not run for long, it was a spontaneous outburst of activity which usually faded out fairly quickly, but nevertheless generated excitement.

Figure 6.24 Several players ran as either a defensive or an offensive tactic, and their behaviour sometimes provoked others to run after them.

Figure 6.25 Two players try to surround an opponent to tag him. In this case they did not succeed; the opponent tagged one of them.

Figure 6.26 Despite having lost GPS coverage, the player still pretends to be part of the game, smiling to others and trying to surprise them.

Participants also tried to surround a person together (figure 6.25). One participant at the Open University pretended to be in the game although he had lost GPS, so he kept walking in the middle of the field (figure 6.26), in a challenging way often approaching people suddenly to scare them.

In a similar manner, another participant who also had lost GPS told his own story:

"I believe that on one occasion where my unit stopped working (no GPS) I continued walking around the field as if I was still in the game. This kept people from the other team looking for 'that last person who still hadn't been tagged".

The same player focused a lot on identifying who was from which team and developed his own 'spying' strategy:

"Whenever I was able to determine if someone was an opponent, I would always try to track their movements through the game and try to catch them off-guard".

In Bristol our participants cooperated a lot with each other. Six people said in the questionnaire that they tried playing the game in pairs in order to rescue each other when tagged. This is evidence that cooperation can evolve through simple game rules and brings to mind our idea of 'linking' (from paragraph 6.2) as a way to become stronger in the game; a concept aiming to prepare the ground for swarms and urban 'flocks'.

In this respect the game play was very different to that of the OU trial, where the exposed and limited space did not encourage strategies to evolve as much. We did observe at the OU however, several cases in which two people would walk along together once they found they were from the same team. Two particular players cooperated a lot, they would untag each other when one of them would get tagged. They often ran together and they tried to surround another player as they were running about. One of them identified teaming up and cornering opponents as a particular strategy that he had tried to employ whenever possible. During the storyboarding process for CitiTag we had developed a scenario where players communicate with others to surround someone. These examples show that such collaborative acts can spontaneously emerge without sophisticated technology, even in an exposed and limited environment. In Bristol, team players tried the strategy of keeping one or two official 'untaggers' away from the 'battlefield', so that tagged players could come back to them. This would not always work as those 'untaggers' sometimes changed place (i.e. the 'base' had to move) or were tagged anyway by the opposing team. We also observed an 'invincible pair': two players who spontaneously teamed up and kept rescuing each other. What is special about this case is that they did not agree on the strategy explicitly, it emerged spontaneously. Here is how one of them described it in the interview:

"The pairing thing happened just like that. We just wandered off together and that was it, we didn't really talk about it. It seemed that it was working and we were constantly releasing each other – we didn't need to say 'let's do this".

We have come to the conclusion that our broad principle of *design for emergence* can be actually translated as *design for pushing boundaries*, based on our observation that when people push boundaries on all fronts, as illustrated in this section, then emergent behaviours occur.

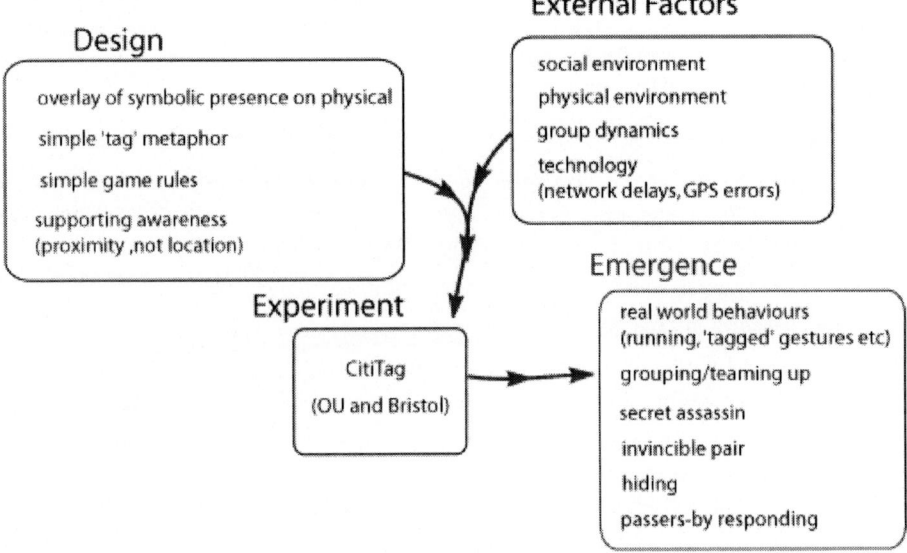

Figure 6.27 Design for emergence in CitiTag

Revisiting our model of design for emergence for mixed reality experiences in chapter 3, we have included our observations from the CitiTag trials in the diagram in figure 6.27, which highlights the emergent behaviours we have just discussed. We have no way of knowing how much of our design (e.g. the simple game rules) and how much of the external factors (e.g. group dynamics) have been influential for certain behaviours to emerge (e.g. grouping/teaming up). But we know that by allowing and encouraging people to experiment and to push boundaries, we can observe a range of emergent behaviours and interesting interactions. In the concluding chapter 7 we discuss how these emergent behaviours could inform the design process (see especially figure 7.2).

6.5.3 Awareness, group belongingness and collaboration

We found that it is possible to associate with the team identity of the Reds or the Greens, even though it so minimal, and feel part of the group, what we broadly named 'group belongingness' or 'team belongingness'. Before running the Bristol trial we played the game with the Mobile Bristol team to test it and we were surprised to find that we wanted our colour to win so much that even our project manager Jo Reid refused to go back for debriefing until the Reds would win at least once!

We asked our participants to rate how important it was for them to know how many players from their team were free and how many had been tagged, in order to investigate the team aspect of the game. In the OU trial most participants found it very important (mean score 80.3), with 7 out of 9 ratings ranging between 70-100 on a 100mm scale ranging from 0 (not at all important) to 100 (very important). Similarly in Bristol, most participants found it also very important (mean score 80.68), with 10 out of 16 replies ranging between 80-100. This result shows that CitiTag is clearly a team game, even if players start playing without knowing who their team mates are and the teams were abstractly defined as 'Reds' and 'Greens'.

We also asked participants to rate how strongly they felt part of their group (Green or Red). In Bristol half of the participants varied between 40-85 with a mean score of 57.5. Some felt part of the Greens or Reds a lot while others less. In Bristol, we had a similar result again with most ratings ranging between 40-100 with a mean score of 62.56. So although our participants thought that it was very important to know the group state (how many were free and how many tagged), they did not associate as strongly with the team identity. For example, in the OU trial, the participant with the lowest rating 12.0 on group belongingness, rated the question on the importance of knowing the team state with a 100. When asked whether they would defect to the opposite team, 10 out of 20 respondents in total would stay loyal to their team, while the other half would join the opposite team to explore game play possibilities or to win. This might also be the result of players changing teams in different sessions they played during the trial, thus team identity not being consistent.

We asked our participants how useful they found the figures on top of the screen displaying the group 'state' (how many players were free and how many had been tagged from each team). At the OU trial, they found the figures very useful for keeping them aware about the state of the group (mean 83.2 with 5 out of 9 ratings between 80-90). Four participants suggested introducing more presence awareness features in their questionnaire responses: being able to see more than one player in the vicinity, directions to the nearest team member, the name of the closest free player to tag and information on how close team mates/enemies are. Similarly in Bristol, participants found the figures very useful for keeping them aware of the group state (mean 76.5 with 12 of 16 people rating them between 70-100). Three participants wanted to see their team members displayed and one other participant wanted to see untagged opponents. These results illustrate that such minimal presence information (numbers of Green and Red free or tagged) can be sufficient to invoke a group state awareness.

For the Bristol trial we found a significant correlation between feeling part of the group (Reds or Greens) and the usefulness of the figures on the top of the device's screen (figure 6.28), used in the game to convey the group state, i.e. how many players of each group were tagged and how many were still free ($r = .794$, $df = 15$, $p < .001$). This indicates that people who felt more part of their group also found this group state awareness information very useful. This result suggests that minimal 'group state' awareness information, along the lines of 'there are 3 free Greens out there', appears to be beneficial for team belongingness and sufficient to evoke the sense of group participation in a shared mixed reality experience.

The feel-good and social cohesiveness effects of being part of a group are emphasized by another finding of the Bristol trial: a significant correlation between feeling part of the group (Reds or Greens) and feeling social (r= .568, df= 15, p<.025).

A particularly interesting finding for our research was a significant correlation found in Bristol between awareness of team members and the importance of knowing the group state (r= .588, df= 14, p<.025). So, people who were more aware of their team members also said it was important for them to know how many players were free and how many had been tagged, the group state, as we can see from the graph in figure 6.29. This shows that players who concentrated on the team aspect of the game also subsequently wanted to have up to date information about the state of each team.

There is an indication that those players who were more aware of their team members, were also 'team players', they cooperated with others and tried to work out strategies. So for example, three players who were highly aware of their team members (96, 87 and 85 respectively), they all mentioned that they tried out cooperation strategies in pairs, going along together to rescue each other. At least four participants in both trials mentioned explicitly that they were trying to free their team mates. For example, a female participant in the OU trial would check the figures and if her team was losing she would try to look around for people to free. Another female participant in Bristol had the same 'team player' attitude: *"I'd look at the numbers, and if the numbers were getting high, I'd go and find them and untag them"*. Some people could even get a 'saviour' status, for example a participant in Bristol commented: *"There was a guy, I think from HP, he was really good cause he actively searched us out and he would come and release us all and then we would run back in"*.

We also found in the Bristol trial that team awareness is significantly correlated to amusement (r= .589, df=14, p<.05), awareness of other people (r=.557, df=14, p<.05) and the importance assigned to winning (r= .659, df= 14, p<.01) illustrated in figure 6.30. The aforementioned three 'team players' who were very aware of their team mates, also thought that winning is very important to them (with ratings of 78, 90 and 100 respectively). We think that the fact that individuals who were keen on winning were also good team players and tried to collaborate is illuminating, as this is not necessarily the case in games in general, where hard competition goes along with individual pursuit of winning. Our findings are also reinforced by another result from the OU trial, a significant correlation between the importance of winning in games and 'group belongingness' (r=.708, df = 8, p <.05). In other words, individuals who said that winning in games was important for them, also felt more strongly part of their team (Greens or the Reds), suggesting a collaborative, team based approach to play.

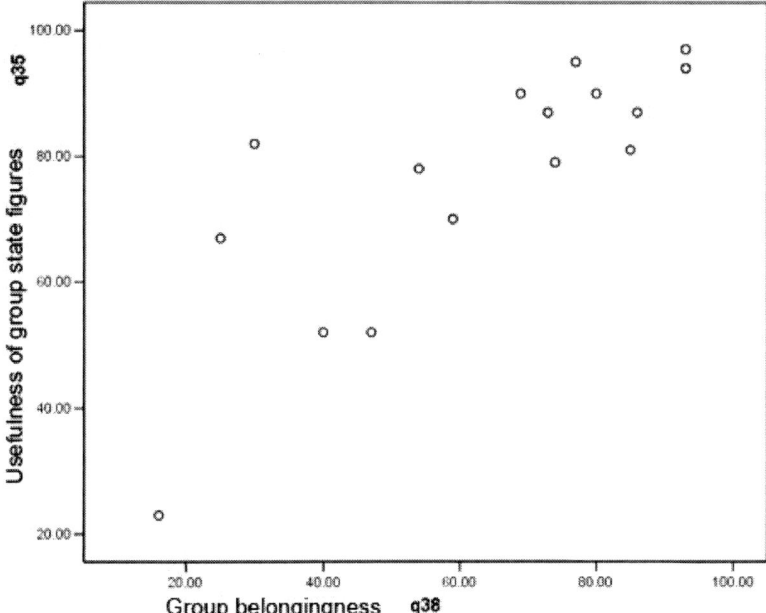

Figure 6.28 Correlation between 'group belongingness' and 'usefulness of figures' in the Bristol trial

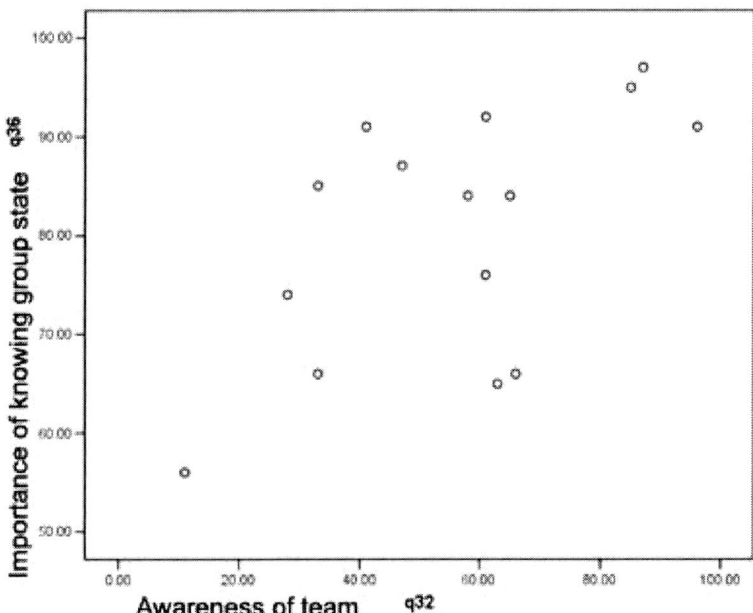

Figure 6.29 Correlation between 'team awareness' and 'importance of knowing group state' in the Bristol trial

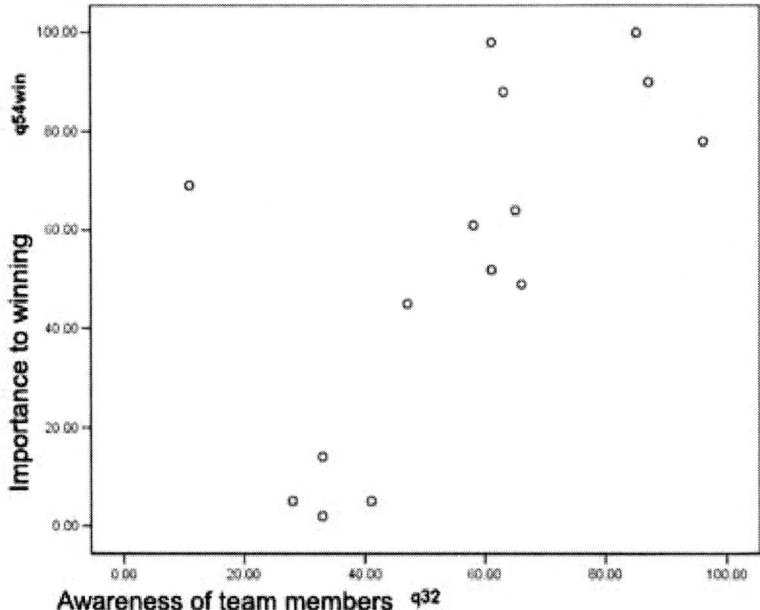

Figure 6.30 Correlation between 'team awareness' and 'importance of winning' in the Bristol trial

To summarise our results on 'team awareness' in simple words, participants who were more aware of their team members were also 1) more amused 2) more aware of other people 3) more keen to win and also 4) they thought it was important to know the group state and 5) they tried to cooperate.

These results allow us to draw conclusions that 'team aware' individuals are good CitiTag players, that CitiTag is a team game and that 'group state' awareness information is important for 'team belongingness' and for cooperation to emerge. They also suggest that such games can be used as teambuilding exercises, for collaborative play to foster interpersonal relationships and social interaction.

6.5.4 Usability and design

Overall there were no major usability problems in both trials and the interface was easy to understand and use. One issue we encountered in Bristol was that the start and end of the game, signified by a message popping up on screen, were not clear enough for all participants, particularly for those who had dropped out of the game for some time because of GPS or wi-fi problems and therefore missed the message. Sometimes the 'game started' message would pop up more than once and that was a bit confusing too at first. These problems were not so evident at the OU trial where everyone could see each other, talk over distance and we synchronised the start and end of each game with more traditional 'coaching' methods (whistle, shouting etc). However, these were all start-up overheads that were quickly out of the way. The main problem our participants in Bristol addressed was that there should be a clearer, more dramatic indication (e.g. a greyed out screen) every time the game had stopped fully working on the device, either because of GPS or wi-fi loss. Although it was possible to see whether one was properly connected by observing the game state (also indicating sustained connection to the game) and GPS numbers indicating positioning at the top of the screen, our participants found this as an obstruction to the game experience. They wanted to not to have to look at the screen too much and worry about whether the game was working on their device or not. Participants reported that they tended to look a lot at the screen (mean 78.2 at the OU and 79.12 in Bristol). Most of them said during the interview that they looked mainly to check whether GPS and wi-fi were working and a few participants mentioned that as time passed they tended to look less and less at the screen. They could keep their thumb at the 'tag' button, so they didn't have to look for that, but instead focus on the environment. Three participants also reported poor visibility because of sunshine.

Another problem reported was a delay in server feedback when the 'tag' or 'rescue' buttons were pressed. Often players were uncertain about whether the button press had gone through so they kept pressing the button continuously. This can be easily solved by removing the button from the interface as soon as it is pressed and displaying some feedback text, e.g. a 'wait…' or 'sending…'. To prevent people pressing the button all the time we could also introduce some very simple rules, for example, if you already tagged a particular person you can't tag them again for x minutes etc.

We found that aural interaction was a very good way to provide awareness and convey the game events. In both trials, participants found the sound alerts in the game very useful (average 76.3 and 62.93 for the OU and Bristol trials respectively). A couple of people at the OU trial said they listened to the others' audio cues to guess what was happening and to figure out who was a friend or foe. A couple of people mentioned during the discussion that listening to sounds was easier than reading text on the screen. We had one sound for all the game alerts in the first trial, but because our participants suggested that different sounds are needed to differentiate the 'states' of the game (e.g. one sound when 'enemy' is in range, another if I got tagged etc) we enhanced this for the Bristol trial. So we used 3 different sounds, one for the 'tag' event, one for getting 'tagged' and one for opportunities to 'untag'. There was also a fourth sound at the end of the game, for the winning team. Bristol participants responded well to those alerts. There were two or three cases with poor or no sound at all, most likely due to a hardware failure. These participants did not find the sound alerts useful of course. The noise in the square made it very hard to listen to the audio cues of other people over distance in order to guess team membership, like in the OU trial. During the interview, participants in Bristol pointed out the advantages of aural interaction, for example as a means to communicate the presence of someone coming closer, like in the film 'Aliens'.

We also found that presence awareness cues are important and the game needs to be enhanced with more presence information and variable proximity alerts. In both trials, participants identified the need for more cues, especially to empower the 'tagged' players to find team mates to untag them. As one participant in Bristol said in the discussion: *'if you were zapping others that was good fun, but once you got tagged, that was it, wasn't it?'* Suggestions included that the interface should indicate where a potential 'saviour' is located and display levels of closeness (hot or cold). Eight participants from both trials (four in OU and four in Bristol) said they wanted to be able to get information for more than one player at a time, team mates and opponents. Two participants from either trials suggested introducing variable levels of proximity. One of them proposed to have two-level proximity awareness and alerts, one for immediately taggable, untaggable players in vicinity and one for players who are very close but not yet in the immediate 'radar' scope (taggable/untaggable). We asked our participants in Bristol in the questionnaire how much they thought they needed a real (functional) radar. More than half of them indicated they needed one (mean 67.00 with 9 out of 16 people ranging between 65-97).

Social trade-offs and decision making need to be incorporated in the game design, following participant feedback. During the group discussion in Bristol, one issue that came up was that there was not enough choice: people did not have to think about whether they should rescue or tag someone. This was automatically resolved by the system with the tagging event having priority over untagging. Some participants would like to have this trade-off and make decisions themselves, for instance to free a team mate rather than tag an opponent if their group number was really low. As one participant said: *'If you could make the choice between tagging or releasing, rather than the system, that would make it more interesting'*. A few participants in both trials also suggested some game design enhancements to empower the 'tagged' players and add variety to the game: a universal 'untagger', an untagging location, shields, 'roving medics', rewards and a time limit. Introducing such variable levels could keep both the novice and experienced players engaged and also encourage people to undertake different roles, leveraging social skills.

Having considered all these issues the key point is to enhance the experience while keeping the game logic and displays as simple as possible, because our trials demonstrated that 'simple' works and it is the right approach to physical, mixed reality games of the kind.

6.6 Conclusions and future work

Overall, our two user trials proved that the basic concept of CitiTag works and showed that playful, rich social interaction can emerge, as well as group behaviours, from a simple game based on *symbolic presence states,* superimposed on real world physical presence. Moreover, these experiences enhance the sense of social participation. Our findings indicate that even minimal 'group state' awareness information (e.g. there are 3 members of my team still free) is sufficient to evoke the social cohesiveness effects of being part of a group (team belongingness). Subsequently, cooperation strategies and team play emerge spontaneously among members of the same team.

Revisiting our design principles in chapter 3, we have come to the conclusion that it is indeed possible to *design for emergence,* as indicated by the range of emergent individual and group actions in our user studies. All these examples were the result of people pushing the boundaries of what was available to them: the game, the technology, the physical environment, the metaphor and the participating players. In this context, design for emergence can be defined as design for pushing boundaries.

The *lightweight design* approach, based on simple rules and presence states, is crucial, ensuring that the device and game interface stimulate real world interaction without adding overheads. This is achieved by *overlaying presence symbolically to create a mixed reality situation.* The fundamental premise is to enhance interaction with others in the real world by adding another layer of reality, so interaction in the form of alerts with sound, as opposed to a rich, immersive computer game experience has proved to be effective for this kind of physical, mixed reality game. Sound cues worked really well in providing awareness and conveying the game events. They were much appreciated by the users and often liberated people from looking at the screen too much. We identified that presence awareness is important, both at the level of the immediate environment (e.g. who is around or near me) but also for the overall state of the game (e.g. how many of my team are free). Our participants felt they needed variable levels of proximity alerts, one for players they could perform actions on (tag, untag) and one for players in vicinity, but not yet in their immediate 'radar' field. So a future version of game needs to be enhanced with more presence information and variable proximity alerts, while still keeping the design as simple as possible.

CitiTag provided *affordances for people to extend the design:* the metaphor of virtual 'tag' motivated the expression of emergent behaviours in the real world, like putting a hand up to indicate that someone has been tagged. Participants could relate to the idea the metaphor conveyed and some also associated the game with other kinds of playground games.

Simplicity is also important to ensure *scalability* and accessibility by a wide range of audiences. The game is potentially scalable and there is no indication from our studies why it could not be played across a whole city if the GPS/wi-fi infrastructure was widely available.

We also identified factors that influence the *experience* of playing CitiTag:
1. Location
2. Group dynamics and the social aspect
3. Using the real world as a game interface
4. The correspondence between the game reality and what actually happens in the real world.

Our participants also indicated that more social trade-offs and decision making need to be introduced in the game design in order to leverage social skills. These observations introduce interesting challenges and open up opportunities for further work in the design of mediated social experiences and games.

Finally, our studies brought to light an additional design implication for future mixed reality applications: to try and introduce, whenever possible, a level of abstraction (e.g. less fine grained presence states) when matching a virtual concept (e.g. a presence state like location or proximity) with the real world user experience. In this way, if, for some reason, the superimposed reality does not correspond with accuracy to what the user really sees or experiences, the breakdown in expectations will not be so palpable and will not hamper the overall experience. For example, by using 'broader' or more abstract levels of proximity, a CitiTag player will not get frustrated if they don't see another player at the exact location they expect them to be due to a GPS error.

While our experiments provided a sense of what it was like to play the game, we would need to run longer term trials in order to find out how this experience could blend with everyday life. We did, however, discuss this with our participants and they identified two ways the game could be played on an everyday basis: spontaneous, as 'turn up in the park and play it' or build a number of tags during the day with a reward system.

Participants in both trials expressed a preference for persistence. One participant suggested that the game should have the form of an event, which goes 'live' for some time at certain locations during the day, maintaining spontaneity and group interaction. Persistent scores and rewards were also mentioned as a desired feature.

Considering then a future version of CitiTag as part of our everyday life, we can imagine it being played on an ad hoc basis, starting when, for example, a critical mass of registered players shares the same location in a public space. Then, they could all receive an invitation to participate and gameplay would emerge, much like the original children's playground tag. In future work, we would like to design and try out an improved version of the game with increasingly larger numbers of people as well as over a longer period of time to uncover more interactions and to understand how these experiences can blend with our everyday life. In the next chapter, we describe the design of a suggested prototype, informed from the results of our studies so far.

Part IV

Reflections on How to Design for Emergence

7. Designing for spontaneous collaborative play based on presence

7.1 Revisiting the research framework

The BumperCar and CitiTag studies have provided insights into the parameters this thesis set out to investigate (figure 7.1), both for online and mixed reality playful environments. Let's look at those parameters across both studies.

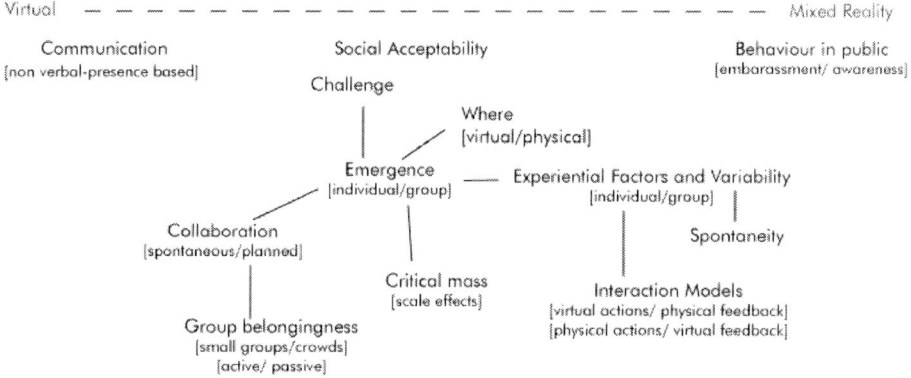

Figure 7.1 Parameters of investigation for collaborative social play based on presence

Communication design is particularly important online where other behavioural cues are absent; the BumperCars experiments explored the minimal, presence-based approach to *communication*. Our findings showed that visual cues were useful for providing context in a playful activity and that signalling based on colour flashes was very easy to miss and therefore not efficient for group coordination. However, we did observe several examples, explained in chapter 5, in which participants managed to collaborate and coordinate their activity just be observing what others were doing, without the need for verbal communication. We also found that it is possible to make assumptions about participants' behaviour expressed visually, without any exchange of verbal communication (e.g. chat or voice). Our conclusion is that when verbal communication is absent, more visual cues about people's presence need to be provided, indicating their state of attention, but also other context-specific presence information (e.g. team membership, intention etc). In the mixed reality studies, visual communication was less important because participants wanted to avoid looking at the screen too much. People needed to be able to maintain an awareness of their surrounding environment and observe others' movement over distance. We found that aural interaction was a very good way to provide awareness and convey the game events. Our participants indicated that presence awareness cues are important and the game needs to be enhanced with more presence information (e.g. team mates) and variable proximity alerts (e.g. people in vicinity but not immediately 'taggable').

In both games, we observed forms of spontaneous 'crowd' behaviour and creative 'rogues', subverting the game activity and introducing new roles and rules. In the BumperCar game, bumping other players out of their parking places resulted in some kind of place swapping activity, reminding us of 'musical chairs', as a whole meta-game with those places emerged. Stealing the facilitator's place was also interesting, challenging the whole notion of an authority in an online experiment, much like misbehaving in a classroom. Participants in these experiments reported that they enjoyed it when they or someone else started 'messing about', which suggests that because of the loosely defined rules and game structure such behaviours were *socially acceptable*. CitiTag participants often ran about in public or shouted to each other over distance during the trial. Although, the trial triggered the curiosity of passers-by in the square, these mob behaviours were socially acceptable and were not disruptive. We did not observe any negative effects of the trial on the social environment. The outbursts of activity were short and the location was quite busy already as people were rushing through, skateboarding and so on. While this location was appropriate for a play, playing CitiTag might have been disruptive in another location. Our participants also felt comfortable *playing this game in public* and not only were they not embarrassed to do so, but moreover they found it acceptable and suitable to play a game of this nature in a public space. It was more important to maintain an awareness of the environment in the Bristol trial than in the Open University trial because of the busy-ness of the location and our findings suggest that we need to provide more support for this overall awareness in future designs.

Next we look into *emergence* and the related parameters in figure 7.1 as our primary point of investigation across both online and mixed reality studies.

7.1.1 Emergence in BumperCars and CitiTag

The online experience of playing BumperCars is very different to the experience of being out in the real world as a CitiTag player, nevertheless we can draw parallels in emergent player behaviour between the two games. In both studies we observed emergence in the form of spontaneous collaboration and creative, 'rogue' behaviours. Let's explore those in detail.

In both BumperCars and CitiTag we had cases of implicit collaboration, coordination that emerged impromptu, either within the game context or beyond it, but most importantly without explicit verbal communication about a specific strategy. In BumperCars we observed the 'victory dance', the spontaneous 'group hug', collaborative colour change rippling and role/space division among members of the same team. In CitiTag players could actually talk to each other and discuss strategies, but usually there was not enough time for such discussions as the game play evolved and certain things emerged without prior coordination. For example, people surrounded individuals together in an attempt to tag them when they identified a person from the opposite team. We also had the case of the 'invincible pair': two players who just went along together and kept rescuing each other without the need to say: 'let's do this'. Similarly, groups were formed in an ad hoc fashion, one person would rescue someone else and then they knew they were from the same team, so they would stay together. Then they would find another person and rescue them and so on, resulting in groups of up to five people moving along together. Participants also demonstrated individual collaborative behaviours: some checked the figures to see how the team was doing and if they noticed that the numbers were getting low, they would try to find tagged team members and untag them. All these are examples of collaborative play. Similarly we observed forms of cheating or misbehaving in both games, which were creative extensions of the game play at the same time. By breaking the rules the 'creative rogues' introduced new dimensions or new roles within the game. In Bumper Cars, the participant who stole the facilitator's place also introduced the meta-game of place swapping, as well as other complex patterns for others to follow. In CitiTag, the 'secret assassin' created an entirely new role in the game by cheating: one can become a 'spy' for their own team and trick the other team. It is also a meta-game with the notion of virtual-physical presence: while being physically among the opposite team, the player belonged virtually to the other team.

The research framework in figure 7.1 indicates that the parameter of collaboration can vary between planned and spontaneous. The studies reinforce this dimension; for instance, the 'group hug' and the 'victory dance' are purely spontaneous and therefore unpredictable: nothing within the game implies or suggests these behaviours, nor do they have a goal or even a symbolic meaning to anyone except to the people who performed them. On the other hand, spontaneous cooperative behaviours like the role/space divisions in the collaborative Pong game, the colour rippling in the group Jam sessions and, in CitiTag, the pairing and teaming up are different. Although collaboration emerges spontaneously, without any higher level control or leadership, the game rules imply or encourage the emergence of those behaviours. For example, participants in BumperCars divided the space to cover territory and defend their side and in CitiTag some people paired up and surrounded lone opponents because they could rescue each other if they were together. In a similar manner, we can identify that some emergent individual behaviours are playful and truly unpredictable (e.g. stealing the facilitator's place) and other emergent behaviours are spontaneous, yet implicitly suggested by the context of the game. Even though no one instructed or suggested to participants to put their hand up when 'tagged', some people did this gesture by associating the CitiTag game with the original tag game, or simply trying to draw attention. Our conclusion is therefore that *emergence* (both individual and collective) can vary between being truly unpredictable and implicit in the game/activity context. Table 7.1 shows a categorization of emergent behaviours in BumperCars and CitiTag along this dimension. Sometimes, the distinction is not entirely clear; for example, one could argue that the behaviour of the 'secret assassin' was completely unexpected, yet it still served the game context, the purpose of tagging opponents while hiding among them.

Table 7.1 Examples of the dimension of emergence between unpredictable and implicit in the game context in BumperCars (B) and CitiTag (C)

Unpredictable emergence	Implicit (strategic) in the context
- Victory dance (B) - Group hug (B) - Stealing facilitator's place (B) - Place swapping (B) - Pretending to be in the game (C) - Interaction with passers-by (C)	- Role/ space division (B) - Colour rippling (B) - Pairing, teaming up (C) - Hand up for 'tagged' (C) - Running, surrounding (C) - Hiding (C) - Secret assassin (C)

Some of the emergent behaviours are active demonstrations of *group belongingness*. The 'group hug' and 'victory dance' examples serve no other purpose but to reinforce the fun, participatory and social aspect our designs have focused on. Although a few CitiTag participants stated strongly that they felt part of their team, more people showed through their actions and comments that they were actively participating. They demonstrated a 'team player' attitude, trying to look for team members to rescue or trying to collaborate with others or simply being more aware of their team members. CitiTag participants who felt more part of their group also found the 'group state' awareness information very useful, suggesting that such minimal information, along the lines of 'there are 3 free Greens out there', can be sufficient to evoke the sense of group participation in a shared mixed reality experience.

The aforementioned emergent behaviours are good examples of what Salen and Zimmerman (2004) describe as *transformative social play:* an instance of play where players use the game context to transform social relationships and where game structures come into question and are re-shaped by player action. Players actively engage with the rule system of the game, manipulating it in order to shift, extend or subvert their relations with other players. In our experiments breaking the rules revealed aspects of 'team belongingness': participants sometimes swapped places in the BumperCar game to get close to other players who performed better. By becoming a spy in CitiTag the 'secret assassin' remained loyal to his team and made it win through trickery.

Salen and Zimmerman explain the phenomenon of transformative social play by highlighting the difference between 'ideal' and 'real' rules. 'Ideal' rules refer to the official regulations in the game (e.g. you can only login to one team in CitiTag and that is your team for the rest of the session) and 'real' rules are the codes and conventions held by a play community (Manning, 1983), a consensus of how the game ought to be played (e.g. bumping around and swapping places). While this is definitely one dimension of transformative social play which helps us understand the impact of 'rogues' within the game, it is not the only one. The aforementioned examples of spontaneous collaborative play also transformed the relationships between players: the players who performed the 'victory dance' or the 'group hug' did not break any particular game rules but extended the meaning of the game by creating their own celebration of team play or simply their own fun activity. Similarly, we observed emergent collaborative behaviours in CitiTag within the game structure and rules, like surrounding a person to tag them and forming groups. But these were also *transformative* behaviours: they extended the game in the physical world, transforming a really simply designed game into a rich real world experience through player interaction. Forming groups and the 'invincible pair' illustrate how players defined and extended their relationships spontaneously without the game enforcing them. Considering that the actual game and dynamics are really simple ('tag'-'untag'), supposedly one could play the game just sitting in their office, having their mobile phone on their desk and other players could come within and out of their proximity range. But what transforms this experience, making it different every time depending on the context where the game is played (for instance, think of the difference between the two CitiTag trials: open field versus urban environment) is the phenomenon of emergence in the real world.

So is emergent collaborative play a kind of transformative social play? Our studies with BumperCars and CitiTag have shown that it is, probably one of the most interesting types, because the social dynamics are unpredictable and can result in different spontaneous behaviours and forms of play every time. Transformative social play also varies the user experience of a game like BumperCars or CitiTag. This in turn suggests a strong link between *emergence* and *experiential variability* in our research framework.

7.1.2 Emergence as experiential variability

Emergence has been discussed extensively in this thesis and our studies revealed that spontaneous individual and group behaviours emerge when people push limits on various fronts. By pushing boundaries people create their own personal experiences of BumperCars/ CitiTag. Therefore, emergence becomes a means of varying the user experience. Conceptualising emergence as *experiential variability* to include our observations from both online and mixed reality play studies, the elements in the diagram (figure 7.2) are our suggested boundaries. Whether designing an online or a mixed reality game, these are fundamental ingredients to provide for exploration, allowing players to extend their meaning and stretch their limitations. These boundaries are the primary elements of the designed experience; it is up to the users to vary this experience through experimentation and play.

The diagram shows a game space in which events take place and which can be explored by players, the player as an individual and as a member of a group (when performing a collective activity). Interactions are influenced by the game rules, which are part of the game space. This is a two way process: the social relationships and group dynamics can also influence the rules, often subverting them or creating new rules. Let's look at each of these elements in detail.

1) Game space

The game space is defined by our design and often by external factors. The BumperCar game has a very closely defined game space allowing expression through colour and movement with collision and yet we saw how collaborative play emerged through these limitations, in such a minimal environment. One could even make assumptions about other people's personality based on their visual behaviour in the game. CitiTag is almost the opposite; although the game itself is really simple, even simpler than the BumperCar, it takes advantage of an open-ended game space, being constantly defined by the context of the location it takes place: how crowded it is, does it have places to hide, how aware of the surroundings do the players need to be, how well the technology works in this location, what is the response of people outside the game and so on. Interactions in CitiTag can be influenced by many external factors. The CitiTag game space is the rich fusion of the virtual with the physical world; therefore when thinking about boundaries, we need to consider both the game design itself and the real world, which is the game interface where collaborative play emerges. *Metaphors* are also part of the game space. In BumperCars players extended the bumping metaphor even in non bumping variants of the game and in CitiTag the 'tag' metaphor encouraged players to make their own associations and introduce the playground 'tag' physicality in the game.

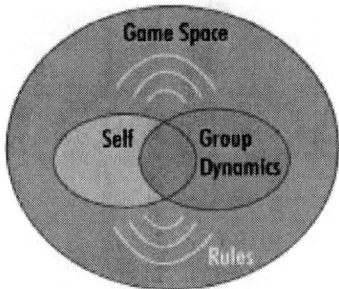

Figure 7.2 When stretching these boundaries, spontaneous collaborative play can emerge.

2) Rules/ authority

Loosely structured rules or a presence of an authority (like the facilitator in BumperCars) can be challenged by the players, introducing emergent play. Stealing the facilitator's place for some time in BumberCars reminded us of naughty schoolboy behaviour and was fun to watch. The exchange of facial expressions undermined the rule of 'no speaking' we tried to introduce in the Open University trial of CitiTag. These are examples of subverting and re-defining the rules through play.

3) Group dynamics and critical mass

All our studies showed that group dynamics need to be accommodated by the game design in a way that differentiates the whole experience. This means that the game should vary depending on the *number of people* participating or if tried with a *different audience* (for instance different age groups). Our experiments with BumperCars showed that even a small increase in the number of participants in these small groups can make a difference in the emerging interaction and result in a variable user experience. For instance, with more participants in group formations and chasing, video analysis revealed that larger teams (three or more), chasing a car of a different colour, were more successful than pairs. Another indication was the difference between the two collaborative Pong experiments: the first with seven participants was more like 'schoolboy football' whereas the second experiment with four participants was much more defensive, like the actual game of Pong. This and other examples suggest that the more people were participating, the more interactions take place among them, so the whole environment appears more lively and full of activity. People also make the experience different every time, individual behaviours emerge and other players respond accordingly, group dynamics constantly defining gameplay. We observed creative rogue behaviours in both games and different groups demonstrated variable behaviours. In the OU CitiTag trial for example, where the user group was more consistent and most people knew each other, participants did not hesitate to talk, discuss strategies openly and even subvert the 'no talking rule'. In Bristol on the other hand, the less consistent user group demonstrated a more strategic, less exposed behaviour and some collaborative acts were implicit, like pairing up. So the more consistent the user group, the more verbal and physical interactions (e.g. running after people, shouting over distance and so on) among participants were observed. In future work, it would be useful to test the same game at the same location and to vary only the participating audience. Group dynamics are unpredictable and this is what is most interesting about multiplayer games in general, so a game that capitalizes on some kind of collaboration can enable these dynamics to have a role and significant impact on the way play evolves.

4) Self (individual role)

Just like groups can define how a certain behaviour (e.g. clustering, swarming) emerges and spreads, individuals should also be able to explore their limits by undertaking their own role in the game. So one can decide to be the good rescuer in CitiTag or the offensive 'tagger' or even play their own game by pretending to be playing the game when they are not. The more possibilities for individual experimentation, the more creative people can be and this is how spontaneous collaboration starts.

Undoubtedly, using the real world and our everyday life context as a game interface is a stimulating experience, because of the physicality of the game space. So far digital games have simulated the real world (e.g. The Sims) or a fantasy world (e.g. RPGs), but with games like CitiTag the reverse starts to happen: the virtual world penetrates the real. A fundamental premise of ubiquitous computing (Weiser, 1996) is that digital technology becomes embedded in everyday artefacts which gain connectivity and intelligence. This creates an additional layer of meaning: objects and individuals can have fictional roles and identities. One of the immediate implications is that people will start using those objects in unexpected ways or behave unpredictably, like in CitiTag, where participants were running around and trying to hide in a public space. Therefore, by blurring the boundaries between the virtual and the physical, a unique opportunity, an experimentation space is created, in which we can study emergence.

What is most fascinating about the emergent behaviours and interactions through a mixed reality game like CitiTag is that, they are only made possible because the virtual counterpart exists. Imagine for instance that a group of adults started playing the normal, physical game of tag in the city centre of Bristol. This would have been fun as an experimental project but it would not be socially acceptable and participants would probably feel embarrassed to run and 'tag' each other in public. The gameplay would rely on the purely physical action of running to get to touch other people. Consider CitiTag, the 'tagging' interaction is mediated through a device and therefore socially acceptable as is now the use of mobile devices in most public spaces outdoors. Further to that, participants introduced the physicality of the game as an extension of their virtual 'state' to the real world: they put their hand up to indicate they got 'tagged', they run briefly after other players and tried to use the environment to their advantage. They also collaborated and went out in pairs to rescue each other; all these physical acts were the results of virtual game events. So virtual events, or better virtual 'states', defined and directed people's behaviour in the physical world: observing that the number of the team 'state' was getting low motivated participants to try and find team mates to rescue, getting 'tagged' made people attempt to draw attention to themselves or go and find people from their own team. This also worked the other way around, people would approach someone they knew was from the opposite team in expectation that the virtual 'tag' event would come up on their screen. Such examples suggest that in the *interaction model* of CitiTag activity flowed from the virtual to the physical and vice versa: virtual actions (e.g. rescue a team member) resulted in physical feedback and interaction (e.g. team up in a pair and move along together) and physical actions (e.g. run after someone) were performed in expectation of the virtual feedback (e.g. the person's name appearing on screen and being able to tag him/her). Although technological failures often hampered this interaction model, our experiments suggest that virtually motivated real world experiences are a promising approach.

We can therefore draw the conclusion that in CitiTag the virtual world acted as a *director of the mixed reality experience*. CitiTag was not a virtual 'tag' game, a mediated alternative to the real world experience of the original playground game, (similar to the Pixeltag concept in chapter 4), but instead it provided symbolic, virtual events that motivated the expression of spontaneity and play in the physical world.

So, is this the end of decades of research efforts to simulate the real world in a virtual environment? Probably not, but it is definitely the beginning of a new direction of research in which the virtual world becomes a 'first class citizen' rather than a limited alternative to a real life experience.

7.1.3   *Emergence feeding back into the design process*

CitiTag is a starting point in a potentially exciting research framework. We know that it is indeed possible to *design for emergence*: by providing just enough context and elements that people can explore (both physical and virtual) unpredictable behaviours can emerge in the real world, motivated by a superimposed fictional reality. But how can these emergent behaviours inform the design of a new game or application? Can emergence inspire design?

In the BumperCar prototype we incorporated the emergent behaviour in the design, by including 'place swapping', something that participants came up with spontaneously, as part of a playful activity online. This proved useful in that case because it served as a fun break for participants and they also got the chance to try to synchronise their moves and colours with different individuals in variable team formations. This example follows the more traditional approach of the iterative design process, to support the observed behaviours by modifying the design appropriately. However, this approach is somehow limited, as not all emergence can be translated into design. Also, what is the added value for the design itself if it supports what people already do spontaneously?

For instance, imagine if we provided a 'cheating' facility for CitiTag players to become 'spies' or 'secret assassins'. It would add an interesting dimension (which we had already considered in our early brainstorms), but we need to look beyond the obvious. The greater challenge is not to change the design to support what people already do, but to open new creative possibilities by understanding how emergent behaviours occur. Therefore, this process of design for emergence is about *identifying the sources of emergent behaviour* in order to understand it and inspire design. These sources can then feed back in the design for emergence model.

For example, by understanding why and how people formed groups in the CitiTag trial we can speculate on what would make them form groups on a larger scale, across the entire city. Our suggested 'linking mechanism' in the storyboards of chapter 6 (e.g. tap to link with someone when in vicinity) only scratches the surface: the point is not to provide people with an interface to form groups, but rather to think about what would motivate people to swarm the city together. A more focused way to conceptualise this is that players should have an advantage against opponents by being in a large group; after all, what we learned from CitiTag is that people teamed up in pairs or groups to be able to rescue each other when tagged and this affected the dynamics of their interactions.

Consider the individual emergent behaviour of the 'secret assassin'; the player experimented with the game rules, in a similar way our creative 'rogue' participant introduced place swapping by stealing other people's parking places or tried to steal the facilitator's place in BumperCars. These examples can inspire experimentation with flexible game rules, introducing elements that can be discovered, used or changed by participants in more than one way. When thinking about the truly unpredictable behaviours, the rationale becomes even more interesting: what made people want to perform a 'group hug' or 'victory dance' in BumperCars? One reason is surely experimentation with the elements of the interface, but there is also the need to show connection to other people when communication channels are restricted or minimized. Social applications that rely on satisfying this need for feeling present, connected with others, can draw inspiration from the experimental and playful design of the BumperCars game activities.

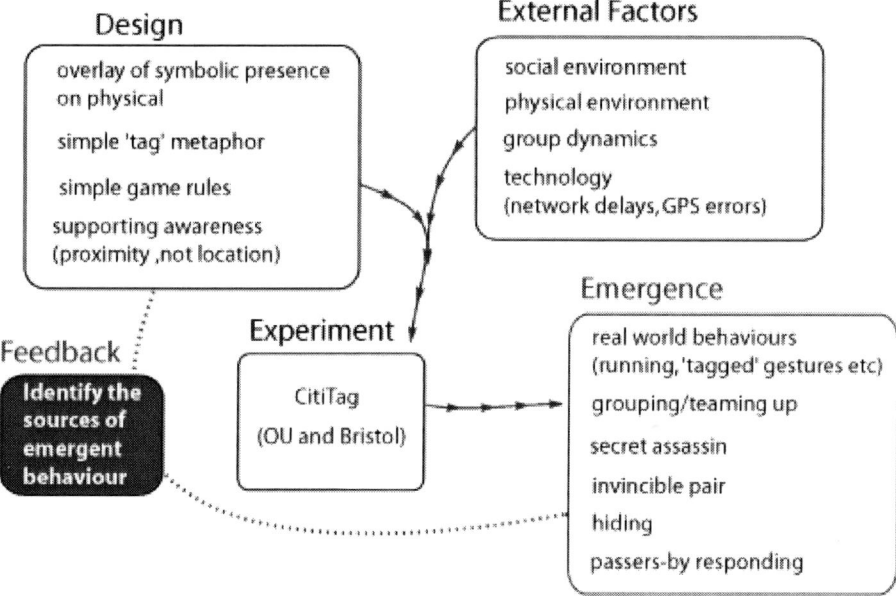

Figure 7.3 The suggested feedback in the design for emergence model in CitiTag: identify the sources of emergent behaviour. For example, being able to rescue each other when close, resulted in small groups of team players getting formed, the invincible pair. The source of the emergent behaviour, the advantage of groups against individual players can be further exploited in the re-iteration of the game towards larger scale.

This is not an easy task though; it seems that design and emergence are two opposite forces, where the former serves to provide an 'engineered' experience and the latter subverts the existing structure, enhancing the experience through the unexpected. Some emergent behaviours can inspire design, while other are just there for us to observe and reflect upon, increasing our design awareness. In order to fully explore our design for emergence model in CitiTag (figure 7.3), it needs to perform another cycle: reflect upon the emergent phenomena, redesign and see what emerges next in another experiment. Our first attempts of design for emergence showed that interesting things can happen and rich interaction can emerge in the real, physical world, fostered through cooperation and creative play. Now we need to review our design with a fresh eye and take it to the next level. It is *emergence reflecting in design*.

7.2 Guidelines for design for emergence for online and ubiquitous multi-user applications

Extending and refining our broad design principles in chapter 3, we have drawn the following guidelines, informed by both our online and mixed reality game studies. These are general guidelines for design for emergence that can be applied to multiplayer games, but are also relevant to social software and other future applications that will involve spontaneous interaction among individuals, mediated or facilitated by communication technologies. They are quite general, because as we mentioned earlier, we can only facilitate emergent interaction by creating an environment where people can explore and push boundaries, but we cannot enforce these phenomena, neither can we assume that there will definitely be some form of spontaneous collaboration.

1) *Provide awareness of the overall state of play.*

Whether online or in the street, it is useful to provide some information about the overall state: the state of the people participating in the same experience. In CitiTag, the group state indicated when the numbers of one team were getting low and motivated players to try and find people from their own team and untag them. Carpenter's synchronised massive game of Pong we discussed in chapter 2 also depended on people being able to see what happens in the whole auditorium on a large overview screen. In our BumperCars experiments with small groups it was important to be aware of the others' activity in order to synchronise or work out a collaborative strategy. Not all activity needs to be necessarily visible; depending on the design goals and limitations this 'group presence' state can be communicated with numbers or symbolic representations or be more explicit if appropriate.

2) *Design for pushing boundaries: include elements that can be explored and extended through experimentation.*

If designing a game, create a flexible, an open-ended game space, allowing people to experiment with different elements of the space (e.g. colour and movement in BumperCars, the physical environment in CitiTag) and use them in variable ways. The rules of play should provide enough structure and context, but at the same time allow people to define the direction and meaning of the game through exploration of strategies and collaboration not defined by the rules. This guideline is in fact more about providing some kind of dynamics within the game, rather than a formal set of rules. For example, CitiTag is based on the simple actions of tagging and untagging in proximity, but the result is unpredictable: the losing team can win in the last minute and alliances can be formed spontaneously without the game imposing them. At large scale (e.g. entire city) unpredictability is even higher, we don't even know whether a game would ever finish. The game play evolves around the struggle of one team to dominate over the other – similarly to the children's playground games where the continuity of the game is ambiguous: some games never finish, others are continued at a later time in another location and so on.

Designing *boundaries to be pushed* means: provide enough context and affordances for people to relate to the experience they are participating in, ensuring at the same time that the experimentation space is not entirely binding, that it allows different interpretations and it creates opportunities for the unpredictable.

3) *Create the need for collaboration without enforcing it.*

Deriving from the above guideline for design for pushing boundaries, this notion follows on a similar tone. We can only create circumstances for individual and group behaviours to emerge, but we cannot enforce or plan emergence. The same holds for spontaneous collaboration, it needs to be facilitated but not enforced. This became clear from our BumperCar studies where we had different cases of teamwork: in the jam colour change session participants were asked explicitly to collaborate in order to create rhythmic colour changes together, but the 'victory dance' and 'group hug' behaviours emerged ad hoc, nobody instructed them and there was no apparent reason for people to perform these acts. They did enjoy creating their own spontaneous collective acts, however, more than the instructed colour jam session or the group formations activity, where again they were asked to form groups and chase other cars together. In CitiTag, a player could play the game individually if they wished and develop their own strategies (e.g. hiding, trying to follow people secretly etc), but people who collaborated benefited much more and enjoyed the game more. Collaboration was not forced or explicitly directed, but evolved during play, so people went around in pairs rescuing each other, formed groups, surrounded others and so on. Reflecting in retrospect on the linking mechanism we suggested in our storyboards in chapter 6, for forming groups in order to become stronger, we now realise that this would not be necessary and it would possibly add another overhead: having to look at your screen to see if your links are still active. Being physically close to someone can be sufficient for groups to be formed ad hoc and this is something that people did spontaneously as they rescued each other.

Our conclusion is that if we provide benefits, some gain from collaboration, people will work out their own collaborative strategies. There is no need for a special interface for this; we can facilitate the process by providing presence features to assist finding potential collaborators. From that point, it is up to the participating individuals to explore the dynamics of collective behaviour, be it movement in urban space or something else that will surprise us.

4) *Lightweight design again: keep it simple.*

This principle keeps coming forward again and again in the thesis, but it is very important as it encompasses our overall approach of using *symbolic presence* and *simple rules* to facilitate emergent interaction. Both the BumperCars and CitiTag studies showed that simple, lightweight design works and that complexity can grow out of simplicity through human interaction, either in the virtual or the physical world.

To summarise, the lightweight design approach for future online and ubiquitous multi-user applications incorporates: *metaphors* that people can relate to, *interfaces* that can be learned quickly by a broad audience, *rules* that can be scaled to involve the participation of large numbers of people and *symbolic presence* for awareness of other people and ongoing activity.

5) *Design experiences that vary with group dynamics.*

We observed that spontaneous collaborative play depends to a large extent on the number of people participating, the dynamics of their interaction and what each individual brings to the group in terms of personal skills, creativity, desire to be a leader and so on. This is true for everyday collaborative activities and our studies are no exception. By designing experiences in which group dynamics play a significant role, more opportunities for emergence are generated: we can run experiments with different audiences or vary the numbers of participants and observe different emergent behaviours. As mentioned above, this is a kind of transformative social play, which allows the social dynamics to play with the rules of the game and define its meaning. This transformative social play enriches the experience of the participating individuals because things can happen beyond the obvious and expected. Our knowledge as designers of interactive systems is also enriched as we gain a deeper understanding of how a particular group of people perceives and extends our design.

There is potentially a very interesting research space in the role of group dynamics in emergent technology-mediated social play that our studies have just initiated: more experiments in the same setting but with different groups would have helped to map this in detail.

The following three guidelines are focused on the field of ubiquitous computing and mixed reality experiences in particular, where a virtual world connects and blurs with the physical.

6) *Use the virtual as a guide to the physical to achieve a balance between fictional reality and real world people, situations, objects.*

It might sound self-explanatory that designing mixed reality applications means creating links between the virtual and the physical, but it is not. In fact, most challenging is *not* defining the links themselves, but doing so in a way that a balance is achieved between the two. It is very common in existing ubiquitous computing applications for people to move around constantly looking at their devices in expectation of certain events, often ignoring real world social conventions and rules (for example conference delegates at UbiComp 2004 stumbled over flowers when playing the Seamful game as they were far too focused on the game play). We believe this balance can be achieved by using the fictional reality as a guide to the physical, demanding from the user to pay attention to the surrounding environment. One such good example from CitiTag is when a player would see that the numbers of his or her team were getting low and start looking around for people to untag. Also, when a player would try to avoid getting tagged and therefore need to observe other people from a distance in order to maintain awareness of those about to enter his or her proximity.

Admittedly, it is indeed a difficult challenge, but it needs to be addressed in order for these applications to succeed in being part of our everyday life.

7) *Abstracting notions of location and proximity.*

Our earlier discussions on affordances, on designing for awareness and interaction with the surrounding social and physical environment are relevant here. We found with CitiTag that the 'mixed reality misconception', when the user experience of the real world does not match his/her expectations based on the information he/she receives on the device, is a significant factor that impedes the whole experience and causes frustration and disappointment. Towards this end, we can anticipate that location positioning technologies and wireless networking will never be perfect, although we might be proved wrong here, but only further away in the future. Somehow this reminds us of other promises, like the one of broadband: the delivery of rich multimedia over the internet has been hanging for quite some time on the expectation that one day bandwidth limits will not be such a major obstacle. But instead web designers learned to work within the limitations and still produce engaging audiovisual content for the web. Similarly, wireless technologies will get better but we cannot expect to have perfect synchronisation in mixed reality applications, just because there are too many things that can go wrong (e.g. server latency, not enough satellites for GPS accuracy, wi-fi drop outs from a passing bus and so on). This is particularly evident in multiuser applications, like CitiTag. So what is the solution? To work within the limitations. As suggested in chapter 6, by abstracting the notion of location and proximity with less explicit visualizations as appropriate, users can still sense the presence of individuals nearby without the need to indicate their exact location or even exactly how far they are. We demonstrate this idea with some sketches for a proposed game prototype in the next section 7.3.

This is a particularly important point. As the language of location-based media is yet to be defined, professionals and researchers use existing traditional conventions, usually geographic maps, for developing location specific applications. For certain cases, like CitiTag, such accuracy is not necessary, so we need to try different approaches in order to find out what works best for each case. Just like the early days of the web, when users had to learn certain conventions, for example that blue underlined text is a link and it can be clicked, in our current early days of location-based applications, a new language has to be established, drawing on existing conventions like maps, but also incorporating, novel, more abstract representation of locality as appropriate.

8) *Persistence, allowing for 'in' and 'out' participation.*

CitiTag participants indicated a preference for persistence while maintaining the spontaneous nature of the game when discussing its potential as part of everyday life. We know well that interactions on the move are short, casual, often interrupted. Consider the situation when while playing a mixed reality game I have to switch off my device and focus on my daily activities. However, I still want to be able to immerse in the experience at a later stage. Possibly the context will have changed by then: I will be in a different location, at a different time and other people participating in the same experience will or will not be around. I should still be able to receive some feedback based on what I was doing earlier in the day and the current situation. Even, if there is no one around to interact with, this feedback can be some information of the overall state of the art, a sense of past events or other people's presence. In this way, I will still feel part of the experience.

To be truly an everyday life experience, a mixed reality application needs to be persistent, allowing the user to 'drop-in' and 'drop-out' in the context of his or her daily activities. The next section suggests some ideas on how we can make the proposed large scale multiplayer location-based game persistent.

7.3 Future work with spontaneous presence-based play

7.3.1 Key research questions and opportunities

Our studies are only a starting point of investigation in collaborative social play mediated through interactive technologies, touching upon the parameters in the research framework, which require further research. One of the fundamental issues with any novel application is whether there is enough motivation and value for the users to adopt it as part of everyday life and use it on a regular basis without the novelty wearing out. Both BumperCars and CitiTag challenge the distinction between time dedicated to play and casual daily activities (e.g. working on the computer, collaborating with others online, shopping in the city centre). A primary incentive for these designs was to see how a playful activity of this kind can blend with the fabric of every day life. We never got to test BumperCars as part of an Instant Messaging system on a regular basis nor CitiTag as part of everyday life activities, because of lack of resources and the typical constraints associated with this kind of research. Another parameter from our research framework that requires further investigation is the one of *critical mass* and the effects associated with increasing the number of participating individuals. If a small increase in the number of participants in small groups can make a difference in the emerging interaction, what would happened if we had tens or hundreds of people participating? In order to find this out we would need a much large user base, so that at different moments there is a critical mass, a sufficient number of players to enable the experience of play. Two promising directions for future research in the area already are: exploring spontaneous presence-based play as part of *daily life* and the dynamics of *larger scale* by increasing the user base. In section 7.3.2 a proposed prototype example, aiming to advance our knowledge from the CitiTag studies is discussed in detail.

A further opportunity for future research lies in the feedback of emergence into the design process. Considering the process of iterative design, we have certainly learned something from the emergent phenomena in CitiTag that can inform the design of a new prototype: mobile technology mediated experiences can give adults some of the sense of fun and spontaneity that are only appropriate in children's playful activities, without any sense of embarrassment or bending unspoken social rules that normally prohibit adults to behave like children. Is this kind of emergent collaborative play different to other forms of escapism, like playing computer games? We believe it is, because a) it uses the real world and real people as a game board and b) it is a simple and accessible form of engagement that brings out the 'inner child' within us in a non violent way. So the next iteration of our design would aim to find out: does this kind of emergence add any value to our everyday experience of the city? Does it make us more sociable?

Ubiquitous computing technologies pose many unknowns, with numerous possible interactions between people, objects, networks and built environments. In this research area, the design for emergence model is particularly useful because it nurtures a culture of exploring the unpredictable. As technology becomes embedded in everyday artefacts, our modes of interaction are constantly re-defined. By observing spontaneous, intuitive uses of technology, we can discover new interaction paradigms as well as ways to satisfy our emotional and social needs. The kind of emergence we explore is also important; in these studies real-time collaborative behaviour (either completely spontaneous or in some way inherit to the game context) and unpredictable bending of the rules were the main forms of emergence observed. But other kinds of emergence could potentially inform the design of new interactive applications. For example, unintentional, asynchronous emergence could occur when people contribute to a collective repository of images or artwork, to form a larger whole pattern over a period of time. The motivation for interactions of this kind that do not have immediate feedback or real-time presence and in which the individual contributes a small part to the entire mass of artwork has not been researched to this date and therefore poses and interesting challenge. By understanding the *sources of emergent behaviour,* designers will be more informed and able to apply the discovered principles in their creative process. In future work we would like to enrich our knowledge based on the design for emergence model, by transferring and adapting the feedback from emergence into the design process of related, yet different products. Consider for instance, the completely unexpected 'group hug' in BumperCars. This kind of interaction could be harnessed in non gaming situations to support participation in groupware applications (e.g. for e-learning, e-working), by appropriating the ability to experiment with the interface and to interact with other people in playful ways without verbal or textual communication. Spontaneous collaborative play online has potential to strengthen community development, by fostering the social bonds among distributed individuals. It can be helpful for online participants to identify good team players and to establish a point of reference for people to meet and get to know each other. There is an opportunity here to leverage social skills that could be applied in the design of future applications for collaborative work, learning, play and social software by providing means of personal expression and group activities.

It would be good for further research using the BumperCars or similar playful online environment to investigate the parameter of group dynamics and *group belongingness*. One aim would be to identify how different kinds of audiences and age groups collaborate and address the challenges posed and whether we observe different emergent behaviours among the variable groups. Our studies with BumperCars showed that people's experience, expectations and individual creativity can influence the emergent interaction, but more focused studies with specific user groups are required to find out how factors like age, gender and previous experiences might affect the group's interaction and coordination. Another aim would be to find out how these participatory experiences can foster group belongingness over a period of time among a community of users. Our participants in the BumperCars studies clearly enjoyed the social factor; generating their own meta-games with others, experimenting, messing about, often resembling children in a schoolyard. But are these experiences sustainable? We can only find out by running larger scale experiments among existing or developing communities, such as groups of distance-learning students who use synchronous online communication media as part of their education practice.

Considering the development and wide application of sensor networks (Marsh, Roussos and Vogiazou, 2004), which enable the communication of sensor-based data online, there are further opportunities to extend the online part of the research in this thesis. Sensors (e.g. motion and orientation sensors, biophysical sensors etc) can be used to communicate a person's presence state online, opening an exciting space for design research in presence-based social gaming. A challenging direction is to study emergent phenomena that are motivated by the visualization and experience of collective physiological states online.

The following section propose directions for future research in mixed reality social play, using an enhanced version of CitiTag for city-wide games as an example. We explore how spontaneous collaborative play could become part of our everyday life and what kind of design would be appropriate, based on what we have found so far.

7.3.2 UrbanSwarm: a proposed example for further research

Consider CitiTag on a large scale, to be played across a whole city and it immediately becomes a new concept, which we call 'UrbanSwarm'. It is a research proposal for a mobile, multiplayer, tag-like game designed to push the limits of scalability and explore emergent, mobile technology-mediated, social play as an everyday life experience. UrbanSwarm is a manifestation of design for emergence: a simply designed mobile game, aimed at encouraging complex, large-scale collective behaviours to emerge spontaneously in the urban environment. With this game we intend to facilitate engaging participatory experiences and social bonding. UrbanSwarm takes a step further from CitiTag to explore the emergence of collective behaviours in the real world on a large scale, with the aim of fostering cooperation, engagement and creativity in our everyday life through the use of simply designed and widely-available technology. Drawing from our findings with CitiTag, we believe that a prototype of Urban Swarm needs to be:

- *Simple* enough to be learned quickly and to be accessible to a casual audience.
- *Scalable* enough to facilitate city-wide play.
- *Collaborative* in order to harness sociability and human communication.
- *Strategic* to require some problem-solving and foster social skills.
- *Flexible* to allow unforeseen, creative behaviours and interactions to emerge.
- *A part of everyday life,* to enable casual 'drop-in' experiences that blend seamlessly with daily activities.

CitiTag studies showed that people enjoyed playing the game in public and being spontaneous. The CitiTag game was suggested as a useful, engaging exercise for team building, which could be part of everyday life, rather than a one-off experience in a specialized venue. UrbanSwarm would aim to further leverage social skills and collaborative play, focusing on the group participation factor to achieve a communal impact.

Here is a scenario of how we envisage interaction through UrbanSwarm to emerge:

You are a commuter walking through Euston station in London and your train has been delayed. You suddenly receive a challenge on your mobile phone informing you that there are 40 Green UrbanSwarmers and 25 Red UrbanSwarmers distributed nearby. You have already signed up as a Red player, so you decide to take up the challenge and participate even though you are outnumbered this time. Using the awareness features on your device and carefully observing the people around you, you manage to find another four Reds. You team up to follow unsuspecting Greens and spread Red around. Then a large group of people emerges from the crowd walking quickly in your direction: they are Greens! Instinctively, you all run to avoid them and rush towards the Underground to 'hide' in the wireless signal-free area. The Green swarm won't follow you there. You relax and chat with the people you just played with, exchanging your UrbanSwarmer details. Two of them are taking the same train as you and it is now time to go!

In this scenario we describe an everyday experience: players can drop-in and drop-out of a game depending on their current activities (e.g. taking advantage of 'dead' waiting time). Notice the emergence: the Reds team up to form a larger group and move altogether. They also use the environment creatively: the London Underground becomes a 'virtual' hiding place, while still being physically visible. The game prototype design will be based on the parameters of strategic team play, presence awareness, large scale and persistence in the context of everyday life.

In CitiTag players formed groups spontaneously without the need to incorporate a group formation mechanism within the game. This was an example of emergent behaviour that was not part of our design and the question stands: would it still emerge on a large scale, if an entire city rather than a confined city square would be the game board? We need to focus on what caused this behaviour rather than simply 'translate' the emergent behaviour in a design specification. The most appropriate approach seems to include some value, a benefit in forming groups. Then people will spontaneously form groups in the real world without the need for a special facility in the game interface like linking. In CitiTag people formed groups because they realized they can rescue each other when tagged, in UrbanSwarm this needs to have a larger effect. So, by clustering and swarming the city together in groups, UrbanSwarm players will acquire 'group power' and become stronger against their opponents. This will probably encourage the formation of large groups: the more people form a group, the more powerful they become and it would be really interesting to see how large the groups can be and for how long would people stay together in a group. In this way, we anticipate that collaborative tactics will evolve spontaneously as players will be interacting in the real world, using their device on-the-move to acquire basic presence awareness information about their team mates and opponents.

Decision making in the game should be challenging enough to leverage social skills and team building, but also simple enough to encourage participation by 'non-gamers' or 'casual gamers'. A visualization of such a simply designed decision making challenge can be for example a meter or progress bar display which increases the power of 'tag' the longer the player waits before tagging an opponent in proximity. If the player tags someone as soon as their device prompts them with a sound alert about the presence of the opponent, then the 'tag' will not last very long and the opponent will get automatically untagged really quickly by a time out event. But if the player waits longer (the meter or bar goes up), then their 'tag' will last longer. Another idea is to use this trade-off of waiting for the last minute before tagging as a means to get shield points that protect the player from future tags. Or both rewards can be combined in one: the longer the player waits the stronger their 'tagging' and they also gain shield. Of course the opponent has the same options. This introduces an interesting prisoner's dilemma in the game as to who will give in first: the anticipation of the moment at which one of the two opponents is going to tag the other. As an alternative to the meter or bar display or in addition to it, this effect can be achieved with increasingly intense sound beeps (like a radar) suggesting that a possible 'tag' encounter is becoming 'hotter'. In this way, a player will not have to look at the screen all the time.

What we suggested earlier as 'group strength' will mean that each player will get more shield points, i.e. more protection against tagging. So the larger and the more sustainable are the groups formed by people swarming the city together, the more powerful and immune to tagging these individuals will become.

CitiTag players said that it would be good to empower the 'tagged' players and a couple of them felt as if the game was over when they got tagged. They suggested various ways to achieve that, like having a universal 'untagger', shields, 'roving medics', rewards and a time limit for being tagged. Such features would keep both the novice and experiences players engaged and possibly encourage interesting social variations (people undertaking different roles). For this reason, UrbanSwarm would allow people to choose between tagging and rescuing. In CitiTag this is done automatically by the system, as a tagging event has priority over an untagging event because the player is always engaged in a coupled, one-to-one interaction; so it would not be fair for them to get tagged while untagging a team member. In UrbanSwarm these interactions will be scaled and multiplied as players will be aware of more people and the vicinity, so that they can choose between tagging or rescuing depending on the state of the team and their strategy. Here we need to be very careful and avoid introducing complexity at all costs: the design of UrbanSwarm should be guided by the key principle of *lightweight design,* keeping it as simple as possible, because our user trials with CitiTag indicated that 'simple' works.

Information on nearby players should be displayed as simply as possible for the mobile context and provide 'just enough' to harness players' imagination, encourage them to interact with people around them and explore the environment in creative ways (e.g. using the London Underground as a 'hiding' place). This is part of our commitment to design for emergence: simple applications to facilitate emergent collective play.

Most CitiTag participants in both trials wanted more presence information about other players, both from their own and the opposite team. Variable proximity alerts were also a desired feature. They proposed two-level proximity awareness and alerts, one for immediately taggable, untaggable players in vicinity and one for players who are very close but not yet in the immediate 'radar' scope (taggable/untaggable). Based on this feedback and in an attempt to scale the concept up to include more players and encourage group interactions, UrbanSwarm can be designed along the lines of the sketches in figures 7.4 and 7.5. Let's have a closer look at these.

The first design aims to abstract the sense of proximity to other players. So the people displayed on the forefront (Nick, George, Debby and Sam in figure 7.4) are the people the player can interact with immediately. So in this screenshot, if the player belongs to the Green team, he or she can either tap on Nick or George to tag one of them or tap on Debby to untag her. The people without names displayed on the other levels, growing smaller and more abstract in the background are the people who are further away, not in the immediate 'radar' of the player. Users will be more aware of others' presence in their neighborhood. This illustration associates the sense of proximity with pictorial conventions of classical art, where objects become smaller and less detailed the further they are supposed to be located from the viewer. The proximity levels can vary and be very approximate depending on the accuracy of the location positioning technology. In this example, we have 10m, 15-20m, 20-50m and entire neighborhood as desired proximity levels. Knowing, however, how frustrating the mismatch of virtual and real presence can be for users as discussed extensively in chapter 6, these distances would probably need to be increased, depending on the accuracy of the implemented technology. So it could be for example that the first level is a mobile network cell (in a dense urban area), the second level is a neighborhood, the third a cluster of neighborhoods and the fourth the entire city. The design would still be the same, therefore inducing enough abstraction of the sense of proximity to ensure that players will not experience the negative effects of 'the mixed reality misconception'.

Let's suppose that there is no one in immediate vicinity to interact with, then the player could communicate with people being further away. He or she could tap on one of the remote players to get their name and send them a quick message or open a voice channel if available. In this way, people can communicate and arrange strategies with others at a distance when there is not much action through a simple to use interface. Additional symbols or numbers can be used to indicate people who are part of a larger group that cannot be displayed because there is not enough space on screen (e.g. +7 to indicate that there are 7 more people close to that person). On the top of the screen people will have a quick overview of the entire game, the states of both teams like in CitiTag: how many people from each team are free and how many have been tagged.

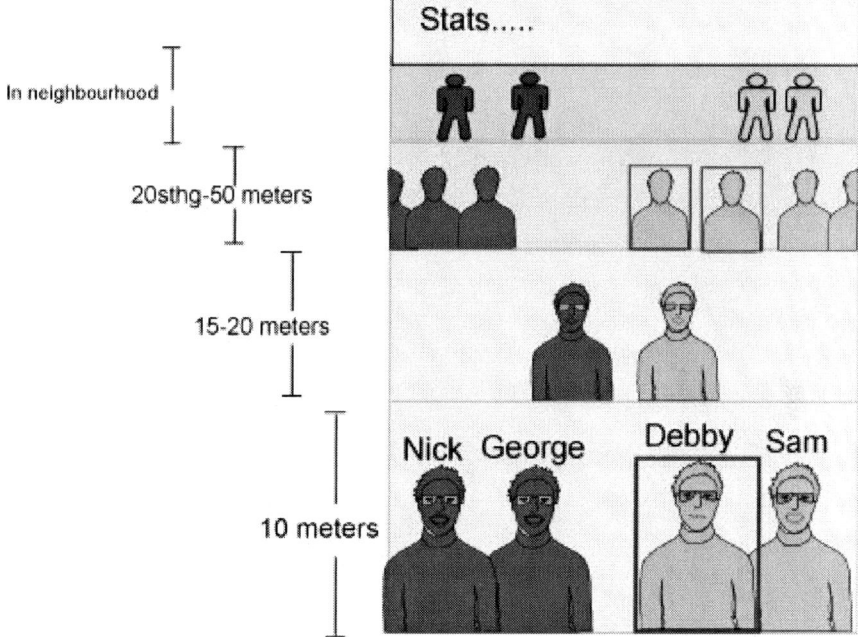

Figure 7.4 Sketch for UrbanSwarm interface displaying players on various proximity levels

In a similar way, yet a bit more complex, the interface in figure 7.5 displays a directional interface, provided the device used (pocket PC or mobile phone) has a compass to determine the user's direction. The arrow on top indicates the direction of movement and moves around the circle as the user moves towards different directions. The proximity levels can be as abstract and as accurate as appropriate in this case too. This is a user-centric display, so the people enclosed by the square in the middle (Yanna, John, Steve and Laura in figure 7.5) are the people in vicinity whether within 10 meters or in the same postcode) with whom the user can interact. The user would be able to see people in further proximity levels by slowly sliding his or her finger over the dots on one of the concentric circles of the radar.

This concept follows our principle in chapter 3 (section 3.2.4) about employing affordances so that people can understand a design. When touching the interface in a circular movement, people will start appearing with their names on the bottom of the screen and just like in the previous design the user can tap on one of them to communicate. Depending on the limitations of the device (e.g. screen space) this interface can be simplified, for example to display only two levels of proximity, i.e. the square in the middle and only one circle with dots and dot clusters around.

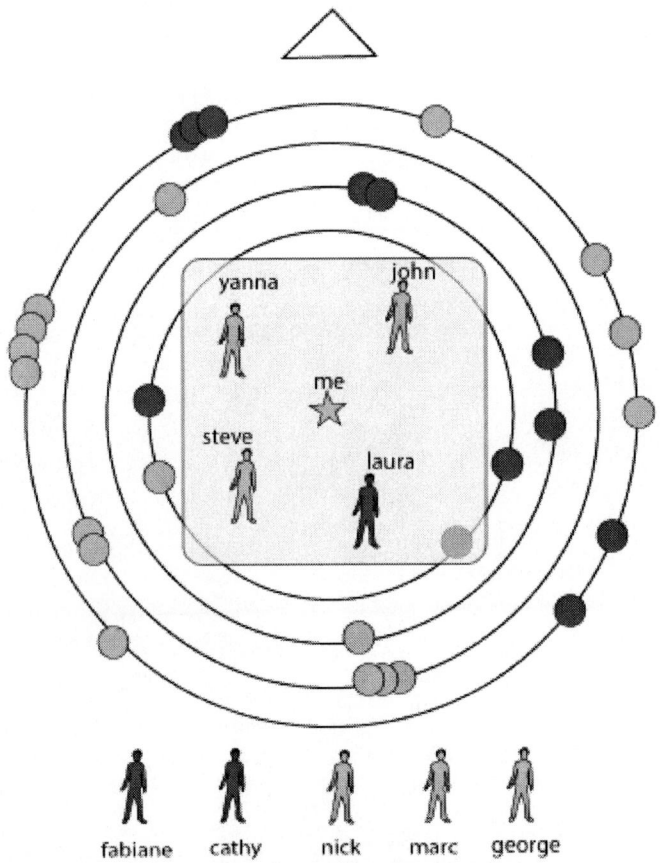

Figure 7.5 Sketch for UrbanSwarm directional interface displaying players in a user-centric radar mode. Players that are very close (e.g. Laura is taggable) are displayed in the centre and by touching along the concentric circles more remote players are displayed in the bottom.

In the broader scope, we want to understand how mixed-reality social experiences can blend with the fabric of our everyday life. In the UrbanSwarm scenario, players can drop in and out of the game, depending on their activity and mood. For this reason, any future prototype should be tried over a longer period of time (weeks) and in the context of casual daily activities.

We would also like to develop scenarios that introduce history and social reputation or rumour in the game. For example, imagine that you are playing

UrbanSwarm and there are no other players around at the moment, but you receive information along the lines of 'Fred was here two minutes ago, he tagged 10 people...' Or alternatively: 'In the past 24 hours 100 greens and 80 reds passed through Euston'. Maybe you could even see a ghost of Fred on your screen. Introducing such features in the game enhances the sense of persistence: even if you are not actively engaged in the game, you can still feel the presence other UrbanSwarm players that were there, so you constantly participate in a virtual experience that is overlaid on the physical world. What kind of implications would this have for our patterns of everyday life? How pervasive should the game be and how much should depend on individual initiative? Will people create social bonds with others they meet through UrbanSwarm? These questions can only be answered if the UrbanSwarm concept is tested through a real prototype.

In the aforementioned UrbanSwarm scenario, the game starts only when a critical mass of registered players are already in the same area. In order to understand and harness the nature of large group spontaneous play we need to test the prototype with as many simultaneous users as possible and over a period of time rather than in one or more organized one-off trials. So what are the limits to scalability? Could UrbanSwarm be played across an entire city? We believe that the scalability challenge can be addressed and the interface designs displayed above already indicate one solution to the limited screen display space. The user can be aware of more people in relevance to their current location as well as of the overall state of the game. But the greatest challenge of scalability lies on the technological side. CitiTag used wi-fi for the multiplayer networking and GPS for location sensing (Quick and Vogiazou, 2004), but this combination has disadvantages. We found that combining GPS with a wi-fi network can be problematic because the two technologies have contradictory requirements: GPS works better in large, open areas without many buildings while wi-fi needs small, confined areas and still will often fail outdoors. Therefore we would need to perform a feasibility study for an UrbanSwarm prototype to decide which of the available technologies is more appropriate. Using the existing GPRS network would certainly be advantageous because it is always available and it can work on the latest mobile phones, which means that we can organize a trial with more people using their personal devices and over a longer period of time. The technological development of such a prototype is very challenging and requires further research.

On a final note, we would like the UrbanSwarm concept to be a starting point for further interdisciplinary research in the design of ubiquitous computing games, crossing the domains of computer science, design and social science. The sketches here propose mobile computing interfaces displaying proximity among individuals. These ideas can be used for the design of location-based, or to be precise proximity-based, social software and other applications that require the collaboration and awareness of people being on the move. A similar interface could be used to represent other, non-physical kinds of proximity: emotional, social and knowledge proximity among individuals. While this proposal is one of several possible approaches for the design of ubiquitous presence-based games (for example aural or haptic interfaces and sensors offer alternative approaches), it is informed by our findings from the CitiTag studies and therefore a suitable framework for further, in-depth research.

7.4 Conclusion: so what?

This thesis has investigated spontaneous, unpredictable uses of technology that are driven by social and collaborative processes, based on our ability to communicate our presence, both virtual and physical, in symbolic ways. By implementing two interactive prototypes, an online and a mixed reality game, we found that collaborative, spontaneous play can enhance the sense of social participation in a group activity. In light of the fact that such social processes and unexpected uses of technology can inspire innovation, we proposed a research model of *design for emergence*, focusing on emergent phenomena as part of an iterative design process. The results from the studies have shown that collective and individual behaviours and creative uses of technology can emerge from a simply designed application, both in the virtual and the physical world. This work suggests directions on how emerging technologies, presence, social dynamics and play can combine in an engaging experience. The focus on emergence has broadened our research angle and it appears to be a useful approach for future work in the area.

Our experiments with CitiTag have brought to light a very interesting perspective: through mixed reality applications virtual elements can direct and enhance casual daily experiences in the context of our physical and social environment in novel ways, with often unexpected results. It is not about imitating reality by providing a virtual alternative, nor about augmenting reality with additional layers of immersive information and visual stimuli on the user's vision sense. The emergent behaviours we observed with CitiTag are personal and collective extensions of the experience in the real world, which would not have existed without the overlay of symbolic presence in the game. Therefore, this thesis finishes with a positive view of ubiquitous social computing as a promising research area: it reverses a long history of attempts to simulate some aspect of the real world – be it a learning experience, collaborative work or play – by actually bringing the virtual world to the forefront as a 'first class citizen', to create new situations and engaging social experiences.

References

1. Abt, C., *Serious Games*. 1970, New York: Viking Press.
2. Allen, C., *Tracing the Evolution of Social Software*, 2004. Available at http://www.lifewithalacrity.com/2004/10/tracing_the_evo.html. Last accessed on 24/01/05.
3. Andersson, N., A. Broberg, A. Bränberg, L.-E. Janlert, E. Jonsson, K. Holmlund, and J. Pettersson, *Emergent Interaction – A Pre-study*2002, Umea, Sweden: UCIT, Department of Computing Studies, Umea University.
4. Arnall, T., *Elastic Space*, 2005. Available at http://www.elasticspace.com/2004/06/mobile-social-software. Last accessed on 24/01/05.
5. Asheron'sCall, 2001. Available at http://zone.msn.com/asheronscall/start.asp. Last accessed on 26/03/2002.
6. Avedon, E. and B.S. Smith, *The Study of Games*. 1971, New York: John Wiley.
7. Baron, J., *Glory and shame: Powerful Psychology in Multiplayer Online Games*, 1999. Available at www.gamasutra.com/features/19991110/Baron_01.htm. Last accessed on 20/12/05.
8. Baron, P., *Location-based mobile phone games*, 2004a. Available at http://www.in-duce.net/archives/locationbased_mobile_phone_games.php. Last accessed on 25/01/05.
9. Baron, P., *Mogi, item hunt*, 2004b. Available at http://www.in-duce.net/archives/mogi_item_hunt.php. Last accessed on 25/01/05.
10. Benford, S.D., M. Fraser, G. Reynard, B. Koleva and A. Drozd. Staging and evaluating public performances as an approach to CVE research. In Proceedings of the *4th international conference on Collaborative virtual environments*. 2002. Bonn, Germany: ACM.
11. Billig, M. and H. Tajfel, Social categorization and similarity in intergroup behaviour. *European Journal of Social Psychology*, 1973. **3**: p. 27-52.
12. Biocca, F., J. Burgoon, C. Harms and M. Stoner. Criteria and scope conditions for a theory and measure of social presence. In Proceedings of the *Presence*. 2001. Philadelphia PA, USA.
13. Birren, F., *Color & Human Response: Aspects of Light and Color Bearing on the Reactions of Living Things and the Welfare of Human Beings*. 1978, NY: John Wiley & Sons.
14. BlastTheory, *Can you see me now*, 2002. Available at http://www.canyouseemenow.co.uk/. Last accessed on 2002.
15. Blinkenlights, *Project Blinkenlights*, 2002. Available at http://www.blinkenlights.de/. Last accessed on 12/06/05.
16. BlueFactory, 2000-2002. Available at http://www.bluefactory.com/web/index.jsp. Last accessed on 26/03/2002.
17. Bluejack, 2003. Available at http://www.bluejackq.com/. Last accessed on 13/01/05.
18. Blythe, M., K. Overbeeke, A. Monk and P. Wright, *Funology : From Usability to Enjoyment*. Human-Computer Interaction Series. 2004: Kluwer Academic Publishers.
19. Boyd, D., *FACETED ID/ENTITY: Managing representation in a digital world*, 2001. MSc Thesis, Massachusetts Institute of Technology.
20. Brown, R., *Group Processes: Dynamics within and between Groups*. 1988, Oxford: Blackwell.
21. Bruce, S.I., *The mobile manhunt*, in *Guardian*. 2002.
22. BuzzoneWebsite, 2004. Available at http://www.buzzone.net/eng/. Last accessed on 24/01/05.
23. Caillois, R., *Man, Play and Games*. 1962, London: Thames an d Hudson. 12.
24. Castronova, E., Virtual Worlds: A First-Hand Account of Market and Society on the Cyberian Frontier. *CESifo Working Paper Series*, 2001. **618**.
25. Chakraborty, R. Presence: A Disruptive Technology. In Proceedings of the *JabberConf*. 2002.
26. Chalmers, M., M. Bell, B. Brown, M. Hall, S. Sherwood, and P. Tennent, *Seamful Game*, 2004b. Available at http://www.seamful.com/. Last accessed on 27/01/05.
27. Chalmers, M., M. Bell, M. Hall, S. Sherwood and P. Tennent. Seamful Games. In Proceedings of the *UbiComp 2004 Demonstrations Proceedings*. 2004.
28. Chalmers, M. and A. Galani. Seamful Interweaving: Heterogeneity in the Design and Theory of Interactive Systems. In Proceedings of the *ACM Designing Interactive Systems (DIS2004)*. 2004.
29. Christiansen, N. and K. Maglaughlin. Crossing from Physical Workplace to Virtual Workspace: be AWARE! In Proceedings of the *International Conference on Human Computer Interaction*. 2003.
30. ConQwest, *Website*, 2005. Available at http://www.conqwest2004.com/. Last accessed on 24/05/05.
31. Costikyan, G., *I Have No Words & I Must Design*, 1994. Available at http://www.costik.com/nowords.html. Last accessed on 11/09/05.
32. Costikyan, G., *Why Online Games Suck (And How to Design Ones That Don't)*, in *The Cursor*. 1998.

33. Crawford, C., Networked Interpersonal Games. *Interactive Entertainment Design*, 1995. **8**.
34. Crawford, C., *The Art of Computer Game Design*, 1997. Available at http://www.vancouver.wsu.edu/fac/peabody/game-book/Coverpage.html. Last accessed on 11/09/05.
35. Cybiko, 1999-2002. Available at http://www.cybiko.com/. Last accessed on 26/03/2002.
36. Danet, B., L. Ruedenberg, B. Gurion and Y. Rosenbaum-Tamari, *'Hmmm Where's that smoke coming from?' Writing play and performance on Internet Relay Chat.*, in *Network and netplay: Virtual groups on the internet*, F. Sudweeks, M. McLaughlin, and S.Rafaeli, Editors. 1998, MIT Press: Cambridge, MA. p. 41-76.
37. Daniel, J.S., *Mega-Universities and Knowledge Media: Technology Strategies for Higher Education.* 1997, London: Kogan Page.
38. Delio, M., *E-Mail Mob Takes Manhattan*, 2003. Available at http://www.wired.com/news/culture/0,1284,59297,00.html. Last accessed on 13/01/05.
39. Desouza, K., Facilitating Tacit Knowledge Exchange. *Comm. ACM*, 2003. **46**(6): p. 85-88.
40. Dey, A.K., G.D. Abowd and D. Salber, A conceptual framework and a toolkit for supporting the rapid prototyping of context-aware applications. *Human-Computer Interaction*, 2001. **16**(2, 3 & 4).
41. Diener, E., Deindividuation, self-awareness, and disinhibition. *Journal of Personality and Social Psychology*, 1979. **37**: p. 1160-71.
42. Diener, E., *Deindividuation:the absence of self-awareness and self regulation in group members*, in *The Psychology of Group Influence*, P. Paulus, Editor. 1980, Lawrence Erlbaum: Hillsdale:NJ.
43. DigitalBridges, 2001. Available at http://www.digitalbridges.com. Last accessed on 26/03/2002.
44. Dodge, M.a. and R. Kitchin, *Mapping Cyberspace*. 2000, London: Routledge.
45. DodgeBall, 2005. Available at http://www.dodgeball.com/. Last accessed on 24/01/05.
46. Donath, J., *Inhabiting the virtual city: The design of social environments for electronic communities*, 1996. PhD Thesis, MIT. http://smg.media.mit.edu/people/Judith/Thesis/
47. Dourish, P., Seeking a Foundation for Context-Aware Computing. *Human-Computer Interaction*, 2001. **16**: p. 229-241.
48. Dourish, P., What We Talk About When We Talk About Context. *Personal and Ubiquitous Computing*, 2004. **8**(1): p. 19-30.
49. Eagle, N., Can Serendipity be planned? *MIT Sloan Management Review*, 2004. **46**(1): p. 4.
50. Eisenstadt, M., *GridMania*, 2000. Available at http://kmi.open.ac.uk/people/marc/gridmania/. Last accessed on 20/05/05.
51. Eisenstadt, M. From Buddy Lists to Buddy Space: Scaleable Experiences of InterPersonal Presence. In Proceedings of the *Presence and Interworking Mobility Summit (PIM2002)*. 2002. Helsinki, Finland.
52. Eisenstadt, M. and J. Komzak, *BuddySpace: Instant Messaging, Maps and Semantics for Enhanced Presence Management*, 2005. Available at http://www.buddyspace.org/. Last accessed on 15/06/05.
53. Eisenstadt, M., J. Komzak and M. Dzbor. Instant messaging + maps = powerful collaboration tools for distance learning. In Proceedings of the *Proceedings of TelEduc03*. 2003. Havana, Cuba.
54. Emilsson, P.K. Presence:Well-done or Medium Rare? In Proceedings of the *Pulver Spring 2001 PIM*. 2001. Boston.
55. Erickson, T., C. Halverson, W. Kellog, M. Laff and T. Wolf, *Social Translucence: Designing Social Infrastructures that Make Collective Activity Visible*, 2002. Available at http://www.pliant.org/personal/Tom_Erickson/. Last accessed on 8/07/02.
56. Ericsson, *Ericsson, Motorola, Nokia and Siemens launch the Mobile Games Interoperability Forum, formerly referred to as the Universal Mobile Games Platform Initiative*, 2001. Available at http://www.ericsson.com/press/20010703-2525.html. Last accessed on Tuesday, July 3 2001.
57. Ericsson, K.A. and H.A. Simon, *Protocol Analysis: Using Verbal Reports as Data*. Second Printing ed. 1996, Cambridge, MA: Oxford University Press.
58. Esato, 2003. Available at http://www.esato.com/web/. Last accessed on 13/01/2005.
59. Espinoza, F.e.a. GeoNotes: Social and Navigational Aspects of Location- Based Information Systems. In Proceedings of the *Ubicomp 2001*. 2001. Atlanta, Georgia.
60. Everquest, 2001. Available at http://everquest.station.sony.com/index.jsp. Last accessed on 26/03/2002.
61. Farkas, I., D. Helbing and T. Vicsek, Mexican Waves in an excitable medium. *Nature*, 2002(419).
62. Felch, J., *Internet-organized 'flash mobs' take strange antics to public places*, in *Denver Post*. 2003: Denver.
63. FIT, F., *Netattack Game*, 2004. Available at http://www.fit.fraunhofer.de/projekte/netattack/index_en.xml?aspect=ar-gaming. Last accessed on 25/01/05.
64. FlashMob, *London Flash Mob Website #2*, 2003. Available at http://www.flashmob.co.uk/mt/2003/08/london_flash_mo_1.php. Last accessed on 13/01/05.
65. Flintham, M., R. Anastasi, S.D. Benford, T. Hemmings, A. Crabtree, C.M. Greenhalgh, T.A. Rodden, N. Tandavanitj, M. Adams, and J. Row-Farr. Where on-line meets on-the-streets: experiences with

mobile mixed reality games. In Proceedings of the *CHI 2003 Conference on Human Factors in Computing Systems*. 2003. Florida: ACM Press.
66. Frank, M.G. and T. Gilovich, The dark side of self and social perception: Black uniforms and aggression in professional sports. *Journal of Personality and Social Psychology*, 1988. **54**: p. 74-85.
67. Freeman, L., Visualizing Social Networks. *Journal of Social Structure, Carnegie Mellon University*, 2000. **1**(1).
68. FriendFinder, *Find Your Friends with Telia FriendFinder*. 2001, Telia.com.
69. Friendster.com, 2002. Available at http://www.friendster.com/. Last accessed on 11/09/05.
70. Garneau, P.-A., *Emergence: Making Games Deeper*, 2002. Available at http://www.pagtech.com/Articles/Emergence.html. Last accessed on.
71. Geirland, J., *UPOC, Inc. is betting that community will be as powerful a concept in the mobile wireless arena as it is for the wired Web.*, in *TheFeature.com*. 2001.
72. Gero, J.S., *Towards a model of exploration in computer-aided design*, in *Formal Design Methods for Computer-Aided Design*, Gero and Tyugu, Editors. 1994, North-Holland: Amsterdam. p. 271–291.
73. Gerrig, R.J., *Experiencing Narrative Worlds*. 1993, New Haven, CT: Yale University Press.
74. Gerrig, R.J. and B.H. Pillow, *Developmental Perspective on the Construction of Disbelief*, in *Believed-In Imaginings: The Narrative Construction of Reality.*, J.D. Rivera and T.R. Sarbin, Editors. 1998, American Psychological Association:: Washington, DC. p. 101-119.
75. Glassner, A., *Some Thoughts on Game Design*, 1997, May. Available at http://www.research.microsoft.com/glassner/work/talks/games.htm. Last accessed on.
76. GoogleMaps, 2004. Available at http://maps.google.com/. Last accessed on 11/09/05.
77. Green, L.N. and E. Bonollo, The Importance of Design Methods to Student Industrial Designers. *Global Journal of Engineering Education*, 2004. **8**(2).
78. Greenberg, S., Context as a dynamic construct. *Human-Computer Interaction*, 2001. **16**(2-4): p. 257–268.
79. Hall, J., *Real and virtual have finally met in fun - reporting on Mogi, the brilliant location-based online multiplayer experience in Tokyo.*, in *TheFeature.com*. 2004.
80. Halverson, C., J. Newswanger, T. Erickson, T. Wolf, W.A. Kellogg, M. Laff, and P. Malkin. World Jam: Supporting Talk Among 50,000+. In Proceedings of the *Poster at the European Conference on Computer-Supported Cooperative Work (ECSCW)*. 2001.
81. Hiler, J., *Artificial Ants*, 2002. Available at http://www.microcontentnews.com/entries/20021220-2589.htm. Last accessed on.
82. Ho-Ching, W.-l., M.K. Inkpen and M. Katherine, *Playing Together: A Taxonomy of Multi-User Video Games*, 2000. Paper, EDGE Lab School of Computing Science, Simon Fraser University, Burnaby, Canada. www.sfu-ca/~fhoching/gi2000_poster_n64.pdf
83. Holland, J., *Emergence*. 1998, Reading, PA: Helix Books. 121-122.
84. Holldobler, B. and E. Wilson., *Journey to the Ants*. 1994, Cambridge, MA: Harvard University Press.
85. Huizinga, J., *Homo Ludens: A Study of the Play Element in Culture*. 1955, Boston: Beacon Press.
86. IJsselsteijn, W. Staying in Touch: Social Presence and Connectedness through Synchronous and Asynchronous Communication Media. In Proceedings of the *HCI International Conference on Human-Computer Interaction*. 2003. New Jersey: Lawrence ErlbaumAssociates.
87. ImaHima.com, 2004. Available at http://www.imahima.com/www/en/welcome/welcome.html. Last accessed on 24/01/05.
88. Isaksen, S., *A Review of Brainstorming Research: Six Critical Issues for Inquiry (Monograph #302)*.1998, NY: Creative Research Unit, Creative Problem Solving Group- Buffalo.
89. ISIT.com, *Scan Mobile to Provide Games and Shopping Over Odigo's Integrated SMS-Instant Messaging Platfrom*, 2001. Available at http://www.isit.com/Press2.cfm?PRid=944&Tech=WIRE. Last accessed on 2002.
90. It'sAlive, *Mobile Games*, 2002. Available at http://www.itsalive.com/. Last accessed on 26/03/2002.
91. Iwatani, Y., *Love: Japanese Style*, in *Wired.com*. 1998.
92. Jacobson, D., On Theorizing Presence. *The Journal of Virtual Environments*, 2002. **6**(1).
93. James, J., *Mobile Gaming: An introduction to the Mobile Gaming Market*2001: Mobile Streams.
94. Jinwoo, K., C. Dongseong and K. Hoyoung, Toward the Construction of Fun Computer Games: Differences in the views of developers and players. *Personal and Ubiquitous Computing*, 1999. **3**(3).
95. Johnson, R.D. and L.L. Downing, Deindividuation and valence of cues: effects on prosocial and antisocial behaviour. *Journal of Personality and Social Psychology*, 1979. **37**: p. 1532-8.
96. Johnson, S., *Emergence: The connected lives of ants, brains, cities and software*. 2001: Penguin Books.
97. Kahn, R. and D. Kellner, New Media and Internet Activism: From the 'Battle of Seattle' to Blogging. *New Media & Society*, 2004. **6**(1): p. 87-95(9).
98. Kellner, S. and F. Petersen, *Plazes*, 2004. Available at http://beta.plazes.com/info/whatis/. Last accessed on 25/01/05.

99. Kelly, K., *Out of Control: The New Biology of Machines, Social Systems and the Economic World.* 1994, Reading, Mass: Addison-Wesley.
100. Knight, W., *Virtual world grows real economy*, in *Newscientist.com.* 2002.
101. Komzak, J. and M. Eisenstadt, *ClustrMaps Beta*, 2005. Available at http://clustrmaps.com/counter/maps.php?url=http://kmi.open.ac.uk/projects/hitmaps/. Last accessed on 20/12/05.
102. Komzak, J. and P. Slavik. Scaleable GIS Data Transmission and Visualisation. In Proceedings of the *7th International Conference on Information Visualisation IV03.* 2003. London: IEEE.
103. Kortuem, G., J. Schneider, J. Suruda, S. Fickas and Z. Segall. When Cyborgs Meet: Building Communities of Cooperating Wearable Agents. In Proceedings of the *Proceedings of the 3rd IEEE International Symposium on Wearable Computers.* 1999b. Washington, DC, USA: IEEE Computer Society.
104. Kortuem, G., Z. Segall and T.G.C. Thompson. Close Encounters: Supporting Mobile Collaboration through Interchange of User Profiles. In Proceedings of the *First International Symposium on Handheld and Ubiquitous Computing (HUC99).* 1999. Karlsruhe, Germany.
105. Kosak, D., *What's This World Coming To?The Future of Massively Multiplayer Games*, 2002. Available at www.gamespy.com/gdc2002/mmog. Last accessed on 5/03/02.
106. Kushner, D., *So What, Exactly, Do online Gamers want?*, in *New York Times.* March 7,2002: NY.
107. Lea, M., R. Spears and D. Groot, Knowing me, knowing you: Anonymity effects on social identity processes within groups. *Personality and Social Psychology Bulletin*, 2001. **27**(5): p. 526-537.
108. Lee, M., *Chatscape: A Behavior-Enhanced Graphical Chat Built on a Versatile Client-Server Architecture*, 2001. partial MSc submission, Electrical Engineering and Computer Science, MIT.
109. Locative.net, *Trans cultural mapping - progress manifesto issues*, 2004. Available at http://locative.net/tcm/workshops/index.cgi?PROGRESS_MANIFESTO_CENTRAL. Last accessed on 25/01/05.
110. Lombard, M., *Presence Explication.*, 2000a. Available at http://nimbus.ocis.temple.edu/~mlombard/Presence/explicat.htm. Last accessed on 30 September 2002.
111. Lombard, M. and T. Ditton, At the Heart of It All: The Concept of Presence. *JCMC*, 1997. **3**(2).
112. Lorenz, K., *On Aggression.* 1967, London: Methuen & Co Ltd.
113. Malone, T., What makes Things Fun to Learn? Heuristics for Designing Instructional Computer Games. *ACM*, 1980.
114. Malone, T.W. Heuristics for designing enjoyable user interfaces: Lessons from computer games. In Proceedings of the *ACM and National Bureau of Standards Conference on Human Factors in Computer Systems.* 1982. Gaithersburg, Maryland.
115. Mamjam, 2003. Available at http://www.mamjam.com. Last accessed on 24/01/05.
116. Mandryk, R. and L.K.M. Inkpen. Supporting Free Play in Ubiquitous Computer Games. In Proceedings of the *Workshop on Ubiquitous Gaming.* 2001. Atlanta, Georgia, USA.
117. Mandryk, R.L., K.M. Inkpen, M. Bilezikjian, S.R. Klemmer and J.A. Landay, *Exploring a New Interaction Paradigm for Collaborating on Handheld Computers*2000: UC Berkeley Technical Report.
118. Manninen, T. Rich Interaction in the Context of Networked Virtual Environments - Experiences Gained from the Multi-player Games Domain. In Proceedings of the *HCI 2001 and IHM 2001 Conference.* 2001: Springer-Verlag.
119. Manning, F., E., ed. *The World of Play.* Proceedings of the 7th Annual Meeting of the Association of the Anthropological Study of Play. 1983, Leisure Press: New York.
120. Maron, M., *The world as a blog*, 2003. Available at http://brainoff.com/geoblog/. Last accessed on 13/01/06.
121. Marsh, A., G. Roussos and Y. Vogiazou. Healthcare Compunetics: An End-to-End Architecture for Self-Care Service Provision. In Proceedings of the *Body Sensor Networks Workshop, Imperial College.* 2004. 6-7 April 2004,London, UK.
122. McGonigal, J. This Is Not a Game: Immersive Aesthetics and Collective Play. In Proceedings of the *DAC.* 2003. Melbourne.
123. McLorinan, A., *The Mobile Games People Play*, 2001. Available at www.ericsson.com.au/about/media_centre. Last accessed on 16/10/03.
124. MeatballWiki, *Social Software*, 2004. Available at http://www.usemod.com/cgi-bin/mb.pl?SocialSoftware. Last accessed on 24/01/05.
125. Mendez, J. and G. Stoll, *Scipionus.com*, 2005. Available at http://www.scipionus.com/. Last accessed on 11/09/05.
126. Meskill, J., *Home of the Social Networking Services Meta List*, 2004. Available at http://socialsoftware.weblogsinc.com/entry/9817137581524458/. Last accessed on 30/01/05.
127. Microsoft, *Asheron's Call*, 2001. Available at http://zone.msn.com/asheronscall/start.asp. Last accessed on 26/03/2002.

128. Milgram, S., *The individual in a social world: essays and experiments*. 1977, Reading,: Mass.: Addison-Wesley Pub Co.
129. Mintz, A., Non-adaptive group behaviour. *The Journal of Abnormal and Social Psychology*, 1951. **46**(2): p. 150-159.
130. Mitchell, W.J., *A computational view of design creativity*, in *Modelling Creativity and Knowledge-Based Creative Design*, Gero and Maher, Editors. 1993, Lawrence Erlbaum. p. 25–42.
131. Mogi, 2003. Available at http://www.mogimogi.com/mogi.php?language=en. Last accessed on 25/01/05.
132. Mudlondon, 2004. Available at http://space.frot.org/mudlondon.html. Last accessed on 25/01/05.
133. Nardi, B., S. Whittaker and E. Bradner. Interaction and Outeraction: Instant Messaging in Action. In Proceedings of the *CSCW'2000*. 2000: ACM Press.
134. NASA, *Astronomy picture of the day*, 2000. Available at http://antwrp.gsfc.nasa.gov/apod/ap001127.html. Last accessed on 20/12/05.
135. Newman, L.S., Intentional and unintentional memory in young children: Remembering vs. playing. *Journal of Experimental Child Psychology*, 1990. **50**: p. 243-258.
136. NokiaGame, 2001. Available at www.nokiagame.com. Last accessed on 26/03/2002.
137. Norman, D., *The design of everyday things*. 2002: Basic Books.
138. Norman, D., *Emotional Design: Why We love (Or Hate) Everyday Things*. 2004: Basic Books.
139. Nova, N., *The impact of Awareness Tools on Mutual Modelling in a Collaborative Game*, October, 2002. MSc, TECFA, University of Geneva, Geneva. http://tecfa.unige.ch/staf/staf-g/nova/msc_nova.pdf
140. Nova, N. and P. Dillenbourg. Impacts of Location-Awareness on Group Collaboration. In Proceedings of the *CSCL SIG Symposium: workshop on Space and Collaboration*. 2004. EPFL, Switzerland.
141. Nova, N. and F. Girardin, *CatchBob!*, 2004. Available at http://craftsrv1.epfl.ch/research/catchbob/. Last accessed on 25/01/05.
142. Odigo, 2002. Available at http://corp.odigo.com/. Last accessed on 26/03/2002.
143. Opie, I. and P. Opie, *Children's Games in Street and Playground*. 1969, Oxford: Clarendon Press.
144. Orkut.com, 2003. Available at https://www.orkut.com/. Last accessed on 11/09/05.
145. Osborne, A.F., *Applied Imagination: Principles and procedures of creative thinking*. 1953, NY: Charles Scribner's Sons. 300-301.
146. Parlett, D., *The Oxford Dictionary of Card Games*. 1992, Oxford: Oxford University Press.
147. Parlett, D., *The Oxford History of Board Games*. 1999, Oxford: Oxford University Press.
148. Paterson, S., M. Zurkow, J. Bleecker and A. Chapman, *PDPal*, 2003. Available at http://www.pdpal.com/. Last accessed on 25/01/05.
149. Paulos, E. Mobile Play: Blogging, Tagging, and Messaging. In Proceedings of the *Ubiquitous Computing*. 2003. Seattle, Washington.
150. Paulos, E. and E. Goodman. The Familiar Stranger: Anxiety, Comfort, and Play in Public Places. In Proceedings of the *ACM SIGCHI*. 2004.
151. PDPal, 2003. Available at http://www.pdpal.com/. Last accessed on 27/05/05.
152. PlayTxt.net, 2004. Available at http://www.playtxt.net. Last accessed on 25/01/05.
153. Preece, J., Y. Rogers and H. Sharp, *Interaction Design: beyond human-computer interaction*. 2002: John Wiley & Sons. 186.
154. Pryor, H. and J. Wood, *GPS drawing*, 2001. Available at http://www.gpsdrawing.com/. Last accessed on 2/05/05.
155. Quick, K. and Y. Vogiazou, *CitiTag Multiplayer Infrastructure*2004, Milton Keynes: The Open University.
156. Raby, F., *Project#26765:Flirt*. 2000, London: RCA.
157. Rafael, V., The Cell Phone and the Crowd: Messianic Politics in Recent Philippine History. *Public Culture*, 2003. **15**(3): p. 399-425(27).
158. Reicher, S.D., Social Influence in the crowd:attitudinal and behavioural effects of deindividuation in conditions of high and low group salience. *British Journal of Social Psychology*, 1984. **23**: p. 341-50.
159. Reid, J., E. Geelhoed, R. Hull, K. Cater and B. Clayton. Parallel Worlds : Immersion in location-based experiences. In Proceedings of the *CHI*. 2005. Portland, Oregon, USA.
160. Reid, J., J. Hyams, K. Shaw and M. Lipson, "Fancy a Schmink?" A Novel Networked Game in a Café. *ACM Computers in Entertainment*, 2004. **2**(3).
161. Remy, M., Wikipedia: The Free Encyclopedia. *Online Information Review*, 2002. **26**(6): p. 434.
162. Rettie, R. Connectedness, Awareness and Social Presence. In Proceedings of the *6th International Presence Workshop*. 2003. Aalborg.
163. Reuters, *Bleep at First Sight*, in *Wired.com*. 1998.
164. Rheingold, H., *Smart Mobs: The Next Social Revolution*. 2002, Cambridge, Mass.: USA: Perseus.
165. Rheingold, H., *Urban Infomatics Breakout*, in *TheFeature.com*. 2004.

166. Rudström, Å., M. Svensson, R. Cöster and K. Höök. MobiTip: Using Bluetooth as a Mediator of Social Context. In Proceedings of the *Ubicomp 2004 Adjunct Proceedings (demo)*. 2004. Nottingham, GB.
167. Ryan, M., *Cyberspace, Virtuality and the Text.*, in *Cyberspace Textuality: Computer Technology and Literary Theory*. 1999, Indiana University Press: Bloomington. p. 78-107.
168. Sacher, H. and G. Loudon, Uncovering the New Wireless Interaction Paradigm. *Interactions*, 2002(January + February): p. 17-23.
169. Salen, K. and E. Zimmerman, *Rules of Play: Game Design Fundamentals*. 2004, Cambridge MA: The MIT Press. 300-311, 70-83, 460-470.
170. Sasaki, H., *Color psychology*, 1991. Available at http://www.shibuya.com/garden/colorpsycho.html. Last accessed on 18/01/05.
171. Saunders, R. and J.S. Gero. Designing for Interest and Novelty: Motivating Design Agents. In Proceedings of the *CAAD Futures*. 2001.
172. SCG, *Design Explorations*, 2002. Available at http://www.research.ibm.com/SocialComputing/SCGdesign.html. Last accessed on 05/05/05.
173. Schiffman, R. and R.A. Wicklund, The minimal group paradigm and its minimal psychology. *Theory & Psychology*, 1992. **2**: p. 29-50.
174. Schön, D.A. and G. Wiggins, Kinds of seeing and their functions in designing. *Design Studies*, 1992. **13**: p. 135–156.
175. Scott, P. and M. Eisenstadt, *Exploring telepresence on the Internet: the KMi Stadium Webcast experience*, in *The Knowledge Web*, M. Eisenstadt and T. Vincent, Editors. 2000, Kogan Page: London.
176. Scuka, D., *A Weather-Affected, Massively Multiplayer, Java-Based I-Mode Game.*, 2001. Available at http://www.japaninc.net/print.php?articleID=59. Last accessed on 11/12/05.
177. Shoemaker, G.B.D. Privacy and Awareness in Multiplayer Electronic Games. In Proceedings of the *Western Computer Graphics Symposium*. 2000. Panorama Mountain Village, Canada.
178. SIA, *Situationist International Archives*, 2001. Available at http://www.nothingness.org/SI/. Last accessed on 13/01/05.
179. Silverstone, R. and Z. Sujon, *Summarizing the Urban Tapestries Social Research: Experimenting with Urban Space and ICTs*, 2004. Available at http://research.urbantapestries.net/socialresearch.html. Last accessed on 27/12/05.
180. SimsOnline, 2002. Available at http://thesimsonline.ea.com/home.html. Last accessed on 26/03/2002.
181. Singel, R., *A Disaster Map 'Wiki' Is Born*, in *Wired.com*. 2005.
182. Socialight.net, 2004. Available at http://www.socialight.net/. Last accessed on 24/01/05.
183. Sotamaa, O. All The World's A Botfighter Stage: Notes on Location-based Multi-User Gaming. In Proceedings of the *Computer Games and Digital Cultures*. 2002. Tampere, Finland: Tampere University Press.
184. Stolterman, E. Uninteded Use. In Proceedings of the *Shaping the Network Society, Patterns for Participation,Action, and Change*. 2002. Seattle.
185. Stone, H., J. Sidel, S. Oliver, A. Woolsey and R.C. Singleton, Sensory Evaluation by Quantitative Descriptive Analysis. *Food Technology*, 1974. **Nov 1974**: p. 24-34.
186. Suits, B., *Grasshoper: Games, Life, and Utopia*. 1990, Boston: David R. Godine.
187. Suwa, M., J.S. Gero and T. Purcell, *Unexpected discoveries and S-invention of design requirements: A key to creative designs*, in *Computational Models of Creative Design IV*, Gero and Maher, Editors. 1999, Key Centre of Design Computing and Cognition, The University of Sydney: Sydney. p. 297–320.
188. Tajfel, H., Experiments in intergroup prejudice. *Scientific American*, 1970. **223**: p. 96-102.
189. TelegraphNews, *You have been flash mobbed*, in *Telegraph News Online*. 2003.
190. Terry, M., E.D. Mynatt, K. Ryall and D. Leigh. Social net: using patterns of physical proximity overtime to infer shared interests. In Proceedings of the *CHI'02 Extended Abstracts*. 2002. Minneapolis, Minnesota, USA: ACM Press.
191. Towell, J. and E. Towell, Presence in text-based networked virtual environments or "MUDS". *Presence*, 1997. **6**: p. 590-595.
192. UltimaOnline, 2001. Available at http://www.uo.com/. Last accessed on 26/03/2002.
193. UnwiredFactory, 2001. Available at http://www.unwiredfactory.com. Last accessed on 26/03/2002.
194. UPOC, 2005. Available at http://www.upoc.com/. Last accessed on 20/12/05.
195. UrbanTapestries, 2004. Available at http://urbantapestries.net/. Last accessed on 25/01/05.
196. Viegas, F. and J. Donath. Chat Circles. In Proceedings of the *CHI 99*. 1999. Pittsburgh PA USA.
197. Vogiazou, Y., *Wireless Presence and Instant Messaging*2002: JISC Technology and Standards Watch.
198. Vogiazou, Y. and M. Eisenstadt. Presence Based Play: Towards a Design for Large Group Social Interaction. In Proceedings of the *First International Conference on Appliance Design*. 2003. Bristol, UK.

199. Vogiazou, Y. and M. Eisenstadt, Play based on Presence Awareness: Facilitating Emergent Social Behaviours Online. *International Journal of Interactive Technology and Smart Education,Special Issue on 'Social Learning through Gaming'*, 2005. **2**(2).
200. Vogiazou, Y., M. Eisenstadt, M. Dzbor and J. Komzak. From Buddyspace to CitiTag: Large-scale Symbolic Presence for Community Building and Spontaneous Play. In Proceedings of the *ACM Symposium on Applied Computing*. 2005. Santa Fe, New Mexico.
201. Watson, R.I., Investigation in deindividuation using a cross-cultural survey technique. *Journal of Personality and Social Psychology*, 1973. **25**: p. 342-5.
202. Weiser, M., The Computer for the Twenty-First Century. *Scientific American*, 1991(September): p. 94-10.
203. Weiser, M., *Ubiquitous Computing*, 1996. Available at http://www.ubiq.com/hypertext/weiser/UbiHome.html. Last accessed on 15/05/05.
204. Whitelock, D., D.M. Romano, A. Jelfs and P. Brna, Perfect Presence: What does this mean for the design of virtual learning environments? *Education and Information Technologies*, 2000. **5**(4): p. 277-289.
205. WhoAt, *WhoAt: Meet Someone- Anywhere*, 2004. Available at http://www.whoat.com/go/in/. Last accessed on 24/01/05.
206. Wolfram, S., *A new kind of science*. 2002: Wolfram Media, Inc.
207. Zimbardo, P.G., *The human choice: Individuation, reason, and order versus deindividuation, impulse, and chaos*, in *1969 Nebraska Symposium on Motivation*, W.J. Arnold and D. Levine, Editors. 1970, University of Nebraska Press.: Lincoln, NE. p. 237-307.

Acknowledgements

This PhD thesis has been completed thanks to the contribution and collaborative effort of several people. First of all, I want to express my gratitude towards Professor Marc Eisenstadt: all the four years of PhD research he has been always *present*, either physically or virtually, always inspiring and supportive in every step I set out to do.

I am really grateful to Bas Raijmakers (RCA, STBY), with whom we worked together on the CitiTag project for his ideas and inspiration. I would also like to thank Bas for the wonderful videos he took of the two CitiTag trials and his help in organising and running those trials. Many thanks to his partner Geke Vandijk for her useful suggestions while developing the CitiTag game scenarios.

The CitiTag project was developed jointly by Centre for New Media in the Knowledge Media Institute (KMi) and the Mobile Bristol team at Hewlett Packard Laboratories (HP Labs) in Bristol. A special thanks to Jo Reid, project manager of the Mobile Bristol team for making it happen, her overall support and valuable advice. Erik Geelhoed has been extremely helpful with the analysis of quantitative data from CitiTag trials. I would like to thank the Mobile Bristol team for their help with the implementation of CitiTag and the user trials: Richard Hull, Paul Marsh (HP Labs), Ben Clayton, Tom Melamed (University of Bristol), who provided the Mobile Bristol GPS location-based support for the game, Stuart Martin and John Honniball (University of Bristol) who provided technical support during the trial.

I am very grateful to the team from the Centre for New Media (CNM) in KMi: Kevin Quick and Jon Linney (KMi), who programmed the multiplayer networking capabilities for CitiTag and Peter Scott for overall supervision and support. Chris Valentine from CNM recorded the user trial at the Open University. Lewis McCann, provided outstanding IT support: from setting up the Wi-Fi on a tree with Kevin to carrying out experiments with iPaqs in his frozen garden in order to crack the mystery of Wi-Fi connection loss, Lewis was always there, in both the Bristol and Open University user trials. Thanks to Matt Eanor from KMi who programmed the BumperCar game in Java. Finally, a great thanks to all people who participated in the BumperCar and CitiTag studies for their time and valuable feedback.

I acknowledge funding support from the Open University Knowledge Media Institute and the Greek Scholarship Foundation (IKY), which enabled me to carry out my PhD research with success.